BETWEEN SPECIES

Dear David,

What good fortune we have met.
to explore this lovely work.

Peace,
Cathy

BETWEEN SPECIES

CELEBRATING THE DOLPHIN-HUMAN BOND

EDITED BY

Toni Frohoff, Ph.D., and Brenda Peterson

SIERRA CLUB BOOKS

San Francisco

The Sierra Club, founded in 1892 by John Muir, has devoted itself to the study and protection of the earth's scenic and ecological resources—mountains, wetlands, woodlands, wild shores and rivers, deserts and plains. The publishing program of the Sierra Club offers books to the public as a nonprofit educational service in the hope that they may enlarge the public's understanding of the Club's basic concerns. The point of view expressed in each book, however, does not necessarily represent that of the Club. The Sierra Club has some sixty chapters throughout the United States and Canada. For information about how you may participate in its programs to preserve wilderness and the quality of life, please address inquiries to Sierra Club, 85 Second Street, San Francisco, California 94105, or visit our website at www.sierraclub.org.

Published by Sierra Club Books
85 Second Street, San Francisco, California 94105
www.sierraclub.org/books

Produced and distributed by
University of California Press
Berkeley and Los Angeles, California
University of California Press, Ltd.
London, England
www.ucpress.edu

SIERRA CLUB, SIERRA CLUB BOOKS, and the Sierra Club design logos are registered trademarks of the Sierra Club.

Illustrations on part-opening pages copyright © 2003 by Kate Spencer (www.katespencer.com)
Petroglyph rubbing in "Talking to Beluga" courtesy of Interspecies Inc.

Library of Congress Cataloging-in-Publication Data

Between species: celebrating the dolphin-human bond / edited by Toni
Frohoff and Brenda Peterson.
    p. cm.
   Includes bibliographical references.
   ISBN 1-57805-070-7 (alk. paper)
    1. Dolphins. 2. Human-animal relationships.   I. Frohoff, Toni, 1963–
II. Peterson, Brenda, 1950–
QL737.C432B48 2003
599.53–dc21   2002036685

Printed in the United States of America on paper that is acid-free and totally chlorine-free (TCF). It meets the minimum requirements of ANSI/NISO Z39.48-1992 (R 1997) (Permanence of Paper). ♽

07 06 05 04 03
10 9 8 7 6 5 4 3 2 1

CONTENTS

APPENDICES

ACKNOWLEDGMENTS

T here are many people who have contributed to this project. We are particularly grateful to Chris Stroud and the Whale and Dolphin Conservation Society, the global voice for the protection of cetaceans and their environment, for their generous and invaluable support of this project. Their commitment to raising awareness about cetaceans and to protecting their welfare and survival has been a source of inspiration and support for many of the contributors in this volume.

This book would also not have been possible without the inspiring work of the authors who contributed to it. We are grateful to several people who provided their assistance throughout this project. Leigh Calvez served as an assistant editor, and we thank her for her expert and creative way with words and vision. Tara Kolden, Rebekka Stahl, and Julie Dennis offered invaluable editorial assistance and insightful organization. Jim Frohoff always graciously contributed to this project. The folks at the Whale and Dolphin Conservation Society were always willing and able to assist us in obtaining information for this book, and we thank them for their skills and insights: Erich Hoyt, Dr. Mark Simmonds, Cathy Williamson, Courtney Stark, Vanessa Williams, and Chris Stroud. We are also grateful to Dr. Naomi Rose from the Humane Society of the United States, Mark Berman and Dave Phillips from Earth Island Institute, and the Swiss Working Group for the Protection of Marine Mammals for their invaluable assistance. We thank Philip Hansen Bailey, Charles Richard Lilly, and Roberta Goodman for helping to prepare Dr. John Lilly's manuscript in light of such a major transition. Others who have contributed to this project to whom we are grateful include Wade Doak, Trevor Spradlin, Dr. Lori Marino, Trisha Lamb Feuernstein, Charles Vinick, Dr. Yolanda Alanis Pasini, Dr. Jason Cressey, Dr. Jane Packard, Cynthia Matzke, Angela and Lillsi Zuyovich, "Bear" Nicole, Emily Guftason, Suzie Norris, and Sharon Negri.

Brenda thanks and acknowledges Susan Berta and Howard Garrett of the Whidbey Island Orca Network, Whale Sightings Network, for their daily cetacean-human bond and bulletins of whales journeying along our West

Coast. Many others have been sources of inspiration and support, all of whom could not be listed here.

Toni thanks and acknowledges Michael Harris from Orca Conservancy for his invaluable assistance and Dr. Idelisa Bonnelly de Calventi and the people at the Academia de Ciencias and Bayahibe. She is also appreciative of the accommodations at the Makai Inn on Maui and the Kale Keawe House in Kealakekua Bay, Hawaii.

We give special thanks to our gracious and expert Sierra Club Books editor, Linda Gunnarson, for her artistic vision and guidance and agent Elizabeth Wales, who believed in this book from the beginning. Kate Spencer has kindly provided us with her remarkable cetacean illustrations, and Elisabeth Beller beautifully coordinated the slipstream between art and production. Brenda would like to celebrate her "podmates" and sister cetacean storytellers, Linda Hogan, Sy Montgomery, and Diane Ackerman.

INTRODUCTION

A Bedouin village on the Red Sea encounters a "miracle" dolphin as their new neighbor; two countries come together to transport an orphaned orca home to her family pod in British Columbia; a musician/naturalist researches ancient petroglyphs of cetaceans that inspire modern-day encounters with belugas; and Amazon tribes ceremonially welcome rare pink river dolphins in a shape-shifting dance between species. These are a few of the stories in this international celebration of the complex and intimate bonds between dolphins and humans.

Dolphins and humans have been mutually curious about each other since ancient times. Sometimes we have even saved each other's lives. This kinship has been honored across cultures and continents in myth, art, literature, and science. In the twenty-first century, we have gone beyond our own view of this interspecies connection and begun to ask: What might this bond look like from the dolphins' perspective? To answer this question, and to explore not only the beauty but also the complexity of the dolphin-human bond, we have searched out the very best nature writing, science, and literature.

This groundbreaking anthology builds on the classic 1974 collection *Mind in the Waters*, compiled by Joana McIntyre, which both inspired and documented a movement to protect dolphins and whales at the peak of their peril from fishermen and whalers. Since the 1970s, human interaction with cetaceans (dolphins, porpoises, and whales) has changed radically from an acceptance of hunting cetaceans to watching and protecting them in many parts of the world. Researchers such as Dr. Ken Norris spawned a new field of study—"cetology"—and irrevocably changed cetacean research by studying the animals where they live—in the wild. The concepts of dolphin "consciousness" and "interspecies communication" between dolphins and humans were popularized by scientists such as Dr. John Lilly; artists such as Jim Nollman ("Talking to Beluga"); naturalists such as Wade Doak, author of *Encounters with Whales and Dolphins;* and science-fiction writers such as Douglas Adams, author of the comic intergalactic romp *The Hitchhiker's Guide to the Galaxy.*

Today, with more people seeking encounters with dolphins and whales, as pointed out by naturalist Erich Hoyt ("Toward a New Ethic for Watching Dolphins and Whales"), cetacean-watching activities have become a billion-dollar industry, with more than nine million participants internationally. Although we celebrate a growing appreciation for cetaceans, we remain mindful that many dolphin species face serious challenges to their existence at the same time as they are widely sought for entertainment, recreation, and, to a lesser degree, educational and therapeutic purposes. A surge of commercial programs offering close encounters with dolphins has taken a toll on dolphins over the past two decades—and shows no sign of slowing. Research and management of these programs lag sorely behind their worldwide expansion. The future well-being of dolphins requires that we develop a deeper respect for them as wild animals, not "playthings," and consider carefully the way we affect their lives.

Does our passion for encountering cetaceans have significant effects on their societies, their habitats, and their future? We invited many expert scientists, including Dr. Bernd Würsig and Melany Würsig ("Being with Dolphins") and author/researcher Dr. Kathleen Dudzinski ("Letting Dolphins Speak: Are We Listening?") to take a deep look at the way our human affection can sometimes harm the very animals we long so to meet.

A passion to be in the presence of dolphins has also led to tremendous growth in the controversial business of oceanariums, a lucrative industry that has many critics, including Greenpeace co-founder Dr. Paul Spong ("The Ocean's Chalk Circle") and biologist Dr. Naomi A. Rose ("Sea Change"). The welfare of dolphins in captivity and the removal of individuals from already depleted populations of wild dolphins continue to be serious concerns, as Dr. Rose, Dr. Toni Frohoff ("The Kindred Wild"), and other contributors point out.

With the explosion of "swim with the dolphins" programs worldwide, we asked researchers and therapists to consider this encounter, especially as it affects dolphins. And how, we asked, can we encounter dolphins responsibly? Dr. Betsy Smith ("The Discovery and Development of Dolphin-assisted Therapy"), a pioneer in dolphin-assisted therapy, comes to a surprising conclusion as to the benefits and costs of that therapy—for both humans and dolphins. In another place where dolphins and humans meet, Canadian researcher Cathy Kinsman ("Luminary") follows a solitary, social beluga who persistently attempts to interact with boaters and swimmers, yet whose vulnerable friendliness puts him in constant danger.

We also asked leading ethicists, including Dr. Marc Bekoff ("Troubling Tursiops"), activists such as Ben White ("The Dolphin's Gaze"), and naturalist Jean-Michel Cousteau ("Necessary Kindness") to address the tough questions about our responsibilities and moral dilemmas when facing another species with its own unique culture, communication, and needs. Ever mindful of our human tendency to exploit other animals, Bekoff reminds us that "love is an essential ingredient in the recipe for reconciliation" between species.

Many of the writers in this collection are also asking questions about the future of the dolphin-human bond. If we are to have a twenty-first-century "cross cultural" exchange between species, what are the possibilities for interspecies healing, mutual habitat protection, and learning about each other? British naturalist Dr. Horace Dobbs explores possible healing from afar through guided imagery, storytelling, and high-tech media experiences in the absence of actual dolphins. And, in "Loving with an Open Hand," Joana McIntyre Varawa concludes that perhaps we should leave the dolphins in peace.

At two of the 2001 and 2002 international marine-mammal conferences, cetacean "culture" was one of the most exciting topics of discussion, building upon previously documented evidence of culture among the great apes established by Drs. Jane Goodall, Roger Fouts, and Richard Wrangham, among others. Recently, cetacean researchers Diana Reiss and Lori Marino documented that dolphins—in addition to humans and other great apes—are capable of recognizing themselves in a mirror. This is significant, because most scientists previously assumed that this was a uniquely primate attribute. It also appears that a form of cultural co-evolution has occurred between dolphins and humans, such as the cooperative fishing of indigenous communities on several continents. Naturalist Howard Garrett ("Springer's Homecoming") elaborates on this concept of cetacean culture by reminding us that orcas have language with distinct dialects passed down by their matriarchs; they also have highly cooperative social structures and hunting strategies—all signs of culture we once thought of solely as the province of humans.

Recognizing that other species have complex cultures—that other animals, such as dolphins, are highly sociable, communicative, family-oriented, and keenly intelligent—also means acknowledging that we humans are not the only sentient beings deserving of "rights." The late Dr. John Lilly advances this conversation in what appears to be his last writing on dolphins, "Toward a Cetacean Nation," which we are honored to publish here for the first time. His plea for a "Cetacean Nation" asks for equal rights and respect for cetaceans, whose voices "must be heard." Lilly counsels, "If we think of dolphins as 'lower animals,' then we will not even attempt to meet them as 'equals' worthy of our efforts."

Finally, a word about the scope of this book. We use the word *dolphin* loosely and liberally to describe all whales, dolphins, and porpoises of the suborder Odontoceti (derived from the Greek term for sea animals having teeth). Odontocetes comprise the vast majority of cetaceans—approximately seventy-one of the almost eighty-five species. They occur in three "superfamilies": Delphinoidea (which includes true dolphins, porpoises, and monodontids such as the beluga whale); Ziphoidea (beaked whales); and Physeteroidea (sperm whales). The other suborder, the Mysticeti, is composed of whales with baleen instead of teeth, which, together with the odontocetes, constitute the order Cetacea. In addition to having teeth, the odontocetes are also distinguished from the mysticetes by numerous physical and behavioral characteristics, which include having a single (instead of a double) blowhole, a highly specialized echolocation system, and a pronounced forehead, referred to as a "melon." They are represented by diverse species, including the well-known bottlenose, spinner, and spotted dolphins, the orca or killer whale (which is technically a dolphin), the white beluga whale, and the largest of the toothed whales, the sperm whale.

Because the dolphin-human bond is so diverse in expression, this book gives voice to many perspectives. Certain essays here may seem radical to some and conservative to others. Many contributors express concern about keeping dolphins in captivity. Many communicate their concern about any sociable interaction with cetaceans because of the harm that can come to these animals in the process. Others maintain that, since neither people nor dolphins can be dissuaded completely from interacting with each other, these activities must be better managed for the safety and well-being of the dolphins.

Our deepest concern is that people encounter dolphins responsibly and respectfully. The study of dolphins is a young field, and the writers in this book—all pioneers in their own way—present a variety of recommendations and approaches to being with or near dolphins. Even when specific recommendations are made, however, what is appropriate for one species or location may not be suitable for others. Above all, we hope that this book encourages our readers to ask the question: How do our interactions affect the *dolphins?* In that way, we hope it will inspire all of us to protect dolphins with the same passion with which we seek to interact with them. Perhaps, too, our love for the dolphins can help guide us to a renewed connection with and respect for our water planet.

So please join us, where land and sea mammals meet, on a journey of science and story. We are delighted to share this journey with you.

PART I

Where We Meet:
The Dolphin-Human Bond

The place where dolphins and humans meet can be a place of exquisite joy and empathy. In "Slipstream," nature writer and novelist Brenda Peterson recounts the dramatic turning point in her difficult decision to stop swimming with captive dolphins and devote her life to protecting dolphins in the wild. Anthropologist Ashley Montagu relates how dolphins and humans have been interacting for millennia, providing an overview of dolphins in classic Western mythology in "The History of the Dolphin." Diane Ackerman allows us to enjoy vicariously her encounter with free-ranging spotted dolphins in the Bahamas in the space where, perhaps, the dolphin-human bond is most blissful—an "At-One-Ment" in play with another species.

Having spent countless hours in the field with dolphins, renowned researchers and naturalists Bernd Würsig and Melany Würsig take us to the world's oceans, where they have conducted their highly influential field research. In "Being with Dolphins," we see how inappropriate human interaction with dolphins in their ocean habitat can disrupt entire dolphin populations. In "The Kindred Wild," Toni Frohoff chronicles her personal account of eight wild dolphins captured for a "swim with the dolphins" facility in Mexico and the harsh impact of such programs on dolphins worldwide. In "The Dolphin's Gaze," activist Ben White shares his story of spinner dolphins that inspired his life's work of environmental and animal protection advocacy.

Finally, the late John C. Lilly takes us from the past into the future with an exploration of cetacean intelligence and interspecies communication toward what he calls a "Cetacean Nation." Lilly's chapter represents what appears to be his last writing on dolphins for publication prior to his passing in 2001, following a half-century of pioneering research and writing.

Slipstream

On this familiar, floating wooden dock, my feet are strong and secure in sleek black flippers, my snorkel mask adjusted like a transparent Cyclops eye, my wet suit tight against my body like a second skin. In my stomach, that familiar fluttering, my heart beating excitedly as I prepare to plunge into the chilly saltwater lagoon. It is my winter pilgrimage to a research center in Key Largo, Florida. I visit here, traveling cross-continent from my home in Seattle to meet both my human family and the dolphins we have come to call our "other relatives."

I can name all these dolphins, adolescent females who swim by, dorsal fins rising and falling in an arc that my eyes take in gratefully. Dinghy, Jessica, Samantha of the crooked jaw, and Dreamer glide by and then execute acrobatic leaps and spins in perfect sync. Dreamer is my favorite, with her half-lidded sleepy brown eyes and intimate scrutiny. She cruises up and with a graceful, glossy snout, or rostrum, gently lifts my flippers and legs so I fall backwards on the dock, laughing.

"Now they're ready to play with you, land-lubbers!" the researcher calls down from his perch above the lagoon. He is studying dolphin-human interaction for his doctoral thesis in marine mammal biology and has spent the summer here at Dolphins Plus Marine Mammal Research and Education Center gathering material on the dolphins' altruistic behavior toward children with life-threatening diseases.

In the next lagoon over a little girl with leukemia, so terribly thin her wet suit hangs on her like a sagging blue slicker, floats in the lagoon with a bouncy yellow life preserver. She is all elbows and arms, but no fear. Even through her mask, her astonishingly pale face reveals the wasting disease that will kill her. This time next month she will be dead, but for now all she knows is that this is her heart's desire—to be in the water with a dolphin. And though very weak, she is thrilled, giggling shyly as a dolphin circles, taking her in with a calm, contemplative eye. This twelve-year-old girl is participating in the Make a Wish Foundation, one of several therapeutic programs at this research center. I gaze at her surrounded by dorsal fins, dolphins tenderly lifting her above the water to carry her slowly around the lagoon, as if she were drowning and they her funeral bearers. In ancient Greece it was believed that dolphins carried human souls between our world and the watery realm of the dead. Dolphins were recognized as creatures who go between the worlds, as guides not only of the shipwrecked and drowning but of the living as well.

As I at last slip into the cold winter lagoon, catching my breath, my chest heaving until the wet suit warms the saltwater against my skin, I feel as if I am also going between two worlds: between the exuberance of the past seven years, in which I've made this Key Largo trip to swim with these dolphins, and my present, conflicted emotions over continuing this bond when it means captivity for the dolphins I have come to love. This swim I must make a decision that weighs on my heart like the pressure of a fathoms-deep dive. But for now I can't bear to think that this might be my last swim with these sister creatures.

Once in this lagoon, which borders a bay into the Atlantic Ocean, I forget everything, even myself, for a moment as I listen: yes, the underwater ricochet of dolphin vocalizations. Bleeps, signature whistles, cetacean songs that sound like creaking doors and Geiger counters come closer and closer until suddenly out of the murky green of the lagoon a graceful glide of silver streaks past, one pectoral fin lightly touching my legs. It always amazes me to remember that inside each dolphin's pectoral is a perfect, five-fingered skeletal hand, remnant of their ancestors who returned to the sea as our mammal cousins. To be touched by this hand-fin is to feel chosen. For just as dolphins are conscious breathers and would suffocate if knocked unconscious, these creatures are so agile and synchronize each movement so thoughtfully that even the slightest touch is a conscious choice.

I long ago stopped wondering if a dolphin's memory or intelligence is equal to our own. I have ample evidence and experience that their capacity for flexible thinking, their myriad strategies for learning new behaviors are survival tools my species could find instructive. Dolphin societies have evolved highly sophisticated social structures and communication skills. Their altruism toward their own as well as other species and their astonishing capacity to outwit researchers are what make them so beloved and yet so mysterious.

As a dolphin again streaks by with just the lightest touch to my arm, I wonder which of "the girls"—as my sister and her children call this group of dolphins—has chosen to reach out to me, remembering me from many other swims here. I lift my head above the surface and propel myself above the water with the borrowed power of my flippers. Calling out to the watchful researcher, I ask, "Was that Dreamer who touched me?"

"No," he laughs. "That's the new one. Alphonse."

"But he's so big!" I sputter in surprise and have to take off my face mask and snorkel to clear the fog of my breathing. "He was just a baby . . . "

"Yes, last time you were here. But he's a big boy now."

As Alphonse cruises by, turning his head to catch my eye, my heart is again captured and I am overjoyed to see he has survived so well, grown up in this lagoon with many doting aunts.

Only a year ago I had arrived with my sister Paula and her three daughters in tow for another swim, only to find the lagoon closed because of a birth—Alphonse's long-awaited arrival. Paula has also been devoted to these dolphins ever since they first "midwifed" her third child, Lissy, in 1986. On Paula's first swim with "the girls" she'd been nine months pregnant and the dolphins had forced all the other swimmers out of the lagoon to focus their full attention on my sister. Practiced in echolocation, the dolphins listened to the fetal heartbeat and, as always, found it fascinating. But that day back in 1986, there had been something different, an agitation in the dolphins that prompted the researcher to ask my sister hesitantly, "Is everything all right with your baby?"

As a surgical nurse and the mother of two healthy girls, my sister had satisfied herself that this third birth would go without a hitch. The doctor had warned her, however, after her last delivery that another pregnancy and cesarean section might be life-threatening. Paula listened to the warning of the dolphins' insistent concern and made sure to have blood donors and a specialist on hand for Lissy's birth. As it turned out, Lissy was born jaundiced and with a rare blood incompatibility that required an immediate transfusion of her body's entire supply of blood. In our family, we credit the dolphins and my sister's intuitive skills with saving Lissy's life. And Lissy considers herself half-sister to "the girls."

In 1993, when Lissy was seven years old, I had arrived with my sister's family to find that a dolphin had just been born. This newborn dolphin was for us like coming full circle in the cycle that included human and animal birthing. The research director let us inside because we were old friends. As we tiptoed in, honored to be allowed to see a newborn dolphin, he whispered, "The little one's chances for survival are 50-50. We have hope because this is Dinghy's second calf. The first one died, you know—as so many dolphin newborns do in the wild."

He went on to remind us that in the wild, the many toxins, heavy metals, and PCBs in the oceans are absorbed by cetaceans and stored in their blubber. When a mother gives birth for the first time, her body purges these toxins through her milk and she literally poisons her own calf. "In some places where Eskimos still hunt and eat whales," he said sadly, "the Inuit women are now like cetaceans—when they nurse their young, they risk killing them with all the toxins in their milk. So you see, the way the oceans and the dolphins and whales go will be the way we humans go. It's just a matter of time."

The hope was that Dinghy's lactating purge released most of the toxins from her body into her firstborn calf; without as many pollutants and poisons, this one might survive. "We haven't named him yet," the research director said. "It's bad luck. And we don't want to get too attached, if he also dies."

Even as he spoke, the silver-haired director already showed how deeply attached he was to this newborn. "Look, see that long, deep wrinkle on his little side? That's from being so recently folded in the womb."

Such a tiny creature and so awkward for a dolphin! This newborn, who would one day be named Alphonse, swam right up against his mother, but his breathing had not the practiced glide and arc of a mature cetacean. Instead, as Dinghy surfaced to exhale that "twoosh, twoosh" of misting air at 100 mph, her son belly-flopped and let out a squeak that was more noise than breath. As Dinghy inhaled, then patiently dived, encouraging her calf to do the same with a guiding pectoral fin to his flanks, Alphonse instead sputtered and flailed sideways near the dock.

We all gasped, expecting him to crash into the floating wooden rectangle, but at the last minute one of his aunts glided between the newborn and the dock, gently guiding him back to his mother.

"This is the most important thing a newborn has to learn," the researcher told us. "To plan and synchronize each breath, not only for his own body, but in rhythm with his pod. Can you imagine humans trying to breathe together while moving so swiftly through the water?" he laughed. "I can't even do it for three minutes with one of my own children, much less my wife."

Nor could I imagine the intimacy and awareness it would take to time my own breathing and fluid movements with those of another human being. As I watched Alphonse and his mother pair-bonding and practicing breathing together, his mother twirling and pirouetting protectively near her newborn, I wondered what kind of life this new dolphin would have, if he survived the crucial first six months. Though he was born in captivity, I knew from my countless trips to Dolphins Plus that these dolphins were cared for and even loved, unlike in some aquariums or zoos where they are forced into tedious performances to "earn" their food. The researchers had assured me that these dolphins were let out every week or so into the open ocean and that they willingly returned to the shelter and safety of this lagoon. In this research facility, they were not expected to interact with humans and had "free zones" for retreat from human encounters. They were never fed as a reward for performance or interaction. The dolphins initiated contact, not us. So to be touched or swum alongside, to be eyed or played with was a dolphin's choice.

I also knew that here the dolphins were safe from many perils—from sharks to drift nets to boat parasites and heavy pollutants in the open ocean. They had excellent veterinary care and reliable, healthy food. Nevertheless, captive dolphins often die of sudden, unknown diseases. And the annual infant mortality rate among captive dolphins is no lower than in the wild. Why not?

During my last several swims in Key Largo, I had found myself troubled by the increasing demands that human visitors were placing on these dolphins. Were they stressed? I had seen the research center grow from a shack on the lagoon to a popular, well-run organization offering dolphin swims twice a day. What toll was this taking on the dolphins? A question haunted me: Is it fair to ask the dolphins to give up their freedom so that we can receive the pleasure of their company? What do the dolphins receive from us in return? Is it worth their lifelong captivity?

A tap on the flipper again from a pectoral fin brings me back to the lovely, dark-green depths of the lagoon. Four dolphins, upside down, reveal their pale white and pink bellies as they cruise in a blur below me. Swimming near me is my friend Laura, who has suffered chronic pain for many years. In her stiff wet suit and snorkel she is shy and awkward in the water, swimming at the edges of the lagoon just as she so often stays outside human circles, held back by what she has called "that other territory of physical pain that seems to separate me so from everyone else." Laura is here to swim with the dolphins, not seeking some miracle cure or healing, but simply to float without the pressure and pain of gravity, to rest in another element that is easy on her body, the benevolent buoyancy of this saltwater. Her hopes of an encounter with a dolphin are muted by her desire not to intrude on them—for more than any of my friends, Laura is loath to invade or demand anything of anybody.

So she floats hesitantly away from the delighted splash and play of those of us in the middle of the lagoon, surrounded by curious dolphins. Samantha with her crooked dolphin smile is pushing my feet now as if I were a favorite float toy, and we skim through the water barely touching the surface. I am propelled through the lagoon so fast that the palm trees and other people all blur into green fronds, black-suited figures in bright life preservers, and turquoise water. Right before we run into the dock, Samantha swerves to miss the wooden rectangle and deposits me at the fence. I am at once exhilarated from the encounter but suddenly aware that here I am again facing my dilemma: Samantha and I cannot race into the open ocean together or explore the fifty-plus miles a day that she is capable of traveling. Nor can we leap together over the fence, because although it is low, she is "psychologically imprisoned," as the animal behaviorists call it. Samantha and all the other dolphins here could easily leap the barricade, but will not leave their pod—or their people. Samantha is held captive here as much by her sense of community and belonging as by the underwater fences.

Troubled by these thoughts, I propel myself off the fencing with a strong thrust of my fins and streak through the lagoon, undulating my hips and legs, my hands straight at my sides in imitation of the dolphins' movements.

But whereas they are graceful, my back is sore from such heavy work after only a few laps around the lagoon. I do not have the flexible, long spine that stretches from neck to tail fluke to slide through the waters. Without tail flukes, with separate legs that can paddle but not propel, with ungainly arms that can pull water but not aerodynamically slip through the waves using currents and my body weight to speed up to 30 mph, I can only crawl as best I can. Added to this indignity is the fact that as a human I splash at both ends, unlike the dolphin, who can enter and exit water almost without a ripple.

Back-paddling now to rest my weary spine, I glance over at Laura, who is clinging to the fencing, as if out of breath. Is she in trouble? Is it a cramp from deep within her pain-wracked body? I turn on my belly and begin to swim over to her, but I am not the first to arrive. By the time I can see Laura through the murky depths and the one eye of my snorkel mask, she is surrounded by six carefully circling dolphins. All of a sudden, their sonar echolocation is very loud as they bounce their sonar off her body and listen to what vibrations come back. This is how dolphins "see" the world, with ultrasound—most of it at a higher frequency than human hearing—in much the same way we have so recently figured out how to use radar in submarines to acoustically "see" canyons and valleys and other ships without our eyes.

Dolphin sonar is so sophisticated that for years the Navy has been studying captive dolphins to learn how better to adapt to an underwater world of sound. Ultrasound, which humans have used to image a fetus or to shatter and dissolve a kidney stone or break up a heart clot, is being studied as "vibrational medicine." Human ultrasound is also being used to open narrow blood vessels to help some heart patients avoid bypass surgery by relaxing artery walls and softening fatty deposits. Medical science is discovering that ultrasound is a less-invasive alternative for treating cataracts and even certain kinds of tumors. And some recent evidence indicates that ultrasound frequencies in the region of 2,000 Hz, high above human hearing, nevertheless have a calming and pleasurable effect on human brain waves, possibly by triggering the release of endorphins. Perhaps this accounts for many people's sensations of ecstasy upon encountering dolphins, particularly in the wild.

Over the years I've come to believe from firsthand experience that it is not so much that the dolphins use ultrasound to heal themselves or us; it is simply a vibrational resonance so sophisticated and strong that it works like a tuning fork. Another tuning fork held near a tuning fork that is struck will begin to vibrate at the exact same pitch as the struck one. Our proximity to dolphins or their echolocation and vocalization might simply encourage our own bodies to resonate with a higher frequency, thus bringing our brain waves into the same Delta or Theta brain waves as the dolphins'. These waves are akin to what we humans experience as meditation or dreaming—but Delta and Theta waves are the dolphins' natural state.

Our immune systems may be triggered by the dolphin sonar, which supports our natural healing. But do we actually have to be next to real live dolphins to experience the same resonance? Or could it happen by listening to audio tapes of dolphin vocalizations, which hospitals have found help women in labor get through birthing with less stress. Or in the future, could we develop three-dimensional IMAX-like sensory-surrounding movies that recreate the same experience as actually swimming with a dolphin in the wild? A kind of virtual healing?

Dr. Horace Dobbs, author of *The Magic of Dolphins*, has researched human depression and dolphins for years; he has combined visualization techniques and audio tapes of dolphin vocalizations to pioneer a therapy he calls the "audio pill." His audio tape *Dolphin Dreamtime* is a combination of dolphin vocalizations, sounds of the sea, the human heartbeat, and a background choir. It seems to placate people suffering from mental illness, depression, and other neurological ills, as well as offering relaxation and meditative support. Dobbs believes that much healing can come about through hydrotherapy, visual hologrammatic images, and audio vocalizations—the combination of which approximates the experience of a dolphin encounter without requiring the capture of live animals. Using such techniques, we can borrow skills from the dolphins without enslaving them or mistaking them for some living form of Prozac.

I keep a respectful distance from Laura and the dolphins swimming in ever closer circles around her so that she is held in the exact eye of their whirlpool. Even underwater I can hear her small murmurs of surprise as suddenly all the dolphins leap above her head—all except one. I struggle to see which dolphin is now performing a pirouette to come belly to belly with Laura in a behavior I have never before witnessed. Vertical alongside Laura, this dolphin opens pectoral fins and tenderly takes my friend in an interspecies embrace. Laura gasps in tiny bubbles leaving her lips. Dreamer clasps Laura to her long, silver body. In slow motion, dolphin and woman twirl together through the water, Dreamer always careful to keep Laura's snorkel above water so she can breathe—which she does now steadily and calmly, as if Dreamer is sharing oxygen with her human body, which had for too long been suffocating. What Dreamer shares in that embrace I will never know. I can only theorize.

Laura swears it was not healing as much as it was animal kindness. "My pain did not disappear," she says afterwards as Dreamer swims her to the dock, lifts her bodily up onto the flat wooden platform, and dives deep. "When that dolphin held me and carried me through the water, I felt whole and a sense of well-being." She pauses and pulls off her snorkel mask, tears streaming from her eyes. "That's it!" she says. "I felt what it was like to be well again, because I was with such a well being."

As Laura steps from the dock onto the shore, her legs collapse under her; one of the researchers helps her walk. Her entire body is trembling, as if still vibrating to another frequency not her own, one borrowed from the dolphins and their buoyant world where the effects of gravity cannot curse a body. She trembles like this for more than two hours and then falls into a deep sleep from which she awakes refreshed.

"The dolphins didn't heal me," she still says to this day, whenever we talk about that rare underwater, interspecies embrace. "I have to do that myself. But they left me with the feeling that anything is possible—even that one day I might be without pain."

I ponder this gift of possibilities that the dolphins bestowed so kindly on my friend Laura the next day when I visit another dolphin research center farther down in the Keys called the Dolphin Research Center.

Here at the research and educational center in Grassy Key, the dolphins are unfamiliar to me, since I've come here only three or four times before. Though the facility is concerned and very committed to its dolphins' care, I have never felt comfortable with its program of rewarding dolphins for interaction with humans—too much like "singing for their supper." But today I am here to observe the work of Dr. David Nathanson, a neuropsychologist who has worked for years with stroke victims, Down's syndrome children, and other children at risk. Using flash cards and interaction with dolphins as a reward for learning, Dr. Nathanson's work shows results in terms of increased learning skills for these disabled children.

"But could this learning increase come about just as well from interaction with domestic animals?" A respected marine mammal biologist and dear friend, Dr. Toni Frohoff, has posed an important question. "Say, like cats or dogs—those therapy animals used by the blind or the disabled in programs such as the Delta Society, which trains domestic animals to help people who are blind or disabled or in the hospital?"

With these questions in mind, I again observe Dr. Nathanson at work with a little Down's syndrome boy who has made his animal-assisted therapy internationally famous. This little boy, Dean Paul Anderson, was never expected to talk, let alone learn. But after five years of working with the dolphins at the Dolphin Research Center, Dean Paul has proven himself somewhat of a prodigy, his symptoms less intense than medical authorities predicted. Is it the dolphins who have made a difference?

According to his mother, Cathy Anderson, the dolphins have made a huge difference in her son's development. "Dean Paul thinks of himself as more dolphin than human," she laughs, watching her son propelled around the lagoon by his favorite dolphin, Santini. "When he draws a picture of his family, some of us have dorsal fins. And there are always several dolphins in the drawing, like brothers and sisters. If you watch my son closely you'll see him exhibit definite dolphin-like behaviors—like he'll turn his head sideways to give you his full attention, the way dolphins, with their 180-degree vision, will do. Sometimes when he doesn't make words, Dean Paul makes dolphin bleeps and vocalizations. He even seems to have his own signature whistle."

There is another disabled child in the lagoon with Dean Paul, a little girl who has delayed motor skills and at the age of five still has not said her first word. After working with the dolphins for several days, she amazes her parents by saying quite distinctly and with perfect diction, "Coca-Cola!"

The exuberant parents are so happy to hear their child's first speech that they immediately insist that Dr. Nathanson let their other, nondisabled child swim in the lagoon. But other people's sessions are scheduled and Dr. Nathanson patiently explains that these sessions are therapeutic only.

"Listen, you," says the suddenly hostile parent. "I've paid a lot of money to come here and get my kid some help. As far as I'm concerned, as long as I'm paying for the time, I *own* these dolphins and I can get anything from them I want!"

Dr. Nathanson sighs and walks calmly back to the lagoon to continue his work with the other disabled children, leaving the girl's parents to fume and mutter. Their furor has blown over, but I am still reeling from the woman's words as if caught up in an inner emotional storm.

I own these dolphins and I can get anything from them I want! Her words spin around me and I feel a wave of nausea deep in my belly. Here is the naked human greed that for years I have tried not to face whenever I have watched dolphins at work and play with humans, even in semicaptivity. The woman had revealed the selfish and dominating demands of humankind. Our species has for too long seen other animals as existing only to serve our needs, believing that animals have no distinct life beyond what they can give to us.

Though the palm trees reveal little breeze and the lagoon is calm, my own body trembles with an inner tumult as if a great wind surges through me, changing my own landscape. This is a turning point in my life. Where there had been uneasiness there is now a decision, where there was ambivalence there is now a new awareness. I realize that I can no longer participate in any encounter with wild animals that keeps them captive. A childhood in the forest has taught me this, but I had forgotten those long-ago lessons in my joy to find connection with this, my other family. I remember that in Hawaii a native friend once told me, "Dolphins are obviously your *aumakua*, the animal you belong to, whether you know it or not. Fortunately, you found out who you are and the animal you're meant to be with. Whom you must protect as a relative and who protects you." I had received the graceful lessons and protection of the dolphins as my interspecies siblings, but what had I done to protect them?

Standing in Grassy Key now, outside the Dolphin Research Center where people pay to have their children "healed" by dolphin encounters, where some parents believe they own another animal whose altruism was simply natural, I make my stand. I can no longer support such entrapment, even if it means I will have to give up my own deep connection to the dolphins in Key Largo. It is my responsibility now to give back to the dolphins, they who have given me so much. It is simple reciprocity—resonance. The tuning fork is struck, and I must respond.

Because dolphins have helped shape my character and my life's work with other animals, now I must try to help shape the future for the dolphins, whether in captivity or in the wild. That night I rent a small motel room next to the Dolphin Research Center. It is a place I've stayed before, right on the water. It offers small kayaks, which I have often paddled out to the lagoon, to float there just outside the fence and watch the dolphins inside.

The night sky features a new moon and Venus is sitting inside the luminous crescent like an intimate visitor. As I paddle the still waters I can already hear the Geiger counter bleeps and ricochets of dolphin vocalizations. Keeping a respectful distance so as not to disturb the dolphins, who have already worked all day long with other humans, I bob in the moonlight, humming low, sometimes randomly singing a nonsense tune.

Suddenly two dolphins swerve from their synchronized leaps and streak toward the fence, skidding to a stop at the last moment. It is this stutter and stop of the dolphins' natural speed at the underwater fencing that finally breaks my heart. The fencing symbolizes the invisible, psychological restraint put on these creatures whose natural exuberance should know no limits—certainly not our own.

As I watch the two dolphins leap around their small saltwater lagoon, I cannot bear to think of giving up my long relationship with the dolphins at Dolphins Plus. I know full well that dolphins will continue to be captured worldwide for human use. I argue with myself: Won't they feel abandoned if people who truly care for them don't keep visiting, keep restoring the bond? Don't I have a responsibility toward them and our interspecies friendship?

At that moment one of the Dolphin Research Center trainers comes out into the night to check on the dolphins. Whistling softly, calling out to several of the dolphins in a fond, singsong voice, the researcher sits on his haunches on the dock and carries on a whimsical conversation with one of the dolphins. I can't hear what he is saying, but the dolphin's pleasure in their bond is evident. These dolphins, like those up at Dolphins Plus in Key Largo, are loved and nurtured. During their long captivity, the dolphins are cared for and given everything they need. All but their real lives, their families, their freedom.

"I'm so sorry," I hear myself whisper now to the two dolphins breathing near me behind their underwater fence. "I'm so sorry you've given up so much. I hope knowing us humans has made up for some of what you lost."

Soon the two dolphins are "talking" to me, opening their mouths to vocalize in return, nodding their long, graceful rostrums as if to say, "Can you stay out late and play?" I do, but my heart is subdued. I stay with these two for almost an hour before the wind whips up from the ocean and my kayak is too unsteady to stay afloat. Besides, it is almost midnight and I need to rest. I know, of course, that these dolphins will not sleep; they will simply shut one eye and, in an adaptation that is so simple, yet so sophisticated, rest one hemisphere of their brain while remaining conscious. Perhaps, I reflect as I paddle back to the motel room, dolphins don't need to sleep because they spend their entire lives in a lucid dream.

The next morning I make my way back up the single-lane road to Dolphins Plus. Though it is a slow and sometimes endless road along these reefs connected like a bright coral necklace for a hundred miles between Miami and Key West, I want the trip to take more time than it does. Before I can even grasp what I am about to do, I arrive at Dolphins Plus in the late afternoon, just in time for another swim with "the girls."

"Welcome back," the director grins, "always room for you."

I have not called ahead as I should have. The programs are always full to overflowing, but I am welcome here. Can I really go through with this—my last swim with these beloved dolphins? Giving up my encounters with the girls feels at once like losing relatives or like abandoning those whose lives and well-being seem intricately connected to my own.

Sure enough, here are Dreamer and Samantha welcoming me back "home." They are eager to engage and Dreamer leaps over my legs before I can even get my flippers pulled on. As I enter the lagoon I try to tell them by body language or telepathy that this is my last time with them, that it is not a rejection but a commitment not just to their group but to all dolphins. Somehow this philosophic rationale seems silly or hollow when greeted with such fondness and familiarity. It feels like telling siblings that I can't ever see them again because I'm off to save the world.

Silly, I tell myself, self-righteous, and above all humorless—quite a sin in the cetacean world. In spite of myself I laugh as Samantha prods at my knees teasingly, as if to ask, "Want to race?"—certain she will win.

As I swim leisurely on the surface, my snorkel mask down to watch when dolphins swim below me, upside down, or right by my side, eye to eye, I try to get some sense of how well these dolphins are doing. Are they depressed or overworked, like my sisters who have each endured grueling years as intensive-care nurses? As I watch the dolphins leap over other swimmers and meet my eye with grace and familiarity, I believe that these dolphins are probably in the best shape of any dolphins I've ever witnessed in captivity. Yet they are still not free to come and go as they please. Like ambassadors from another realm, they are perhaps resigned to teach and translate, ever going between two worlds. Like those who must always speak a second language because they are in exile from their homeland, perhaps these dolphins have somehow even volunteered for the job of helping humans recognize other species' intelligence.

A pang of sorrow strikes as I realize that soon I, too, will be in exile from these dolphins. I swim with a heightened awareness of the lagoon and creatures around me—the way one would memorize a beloved face, landscape, or home for the last time. It is like leaving a second home: I take in the lush lagoon, each dolphin eye holding mine, even the raucous cry of the brown pelican who also makes his home here, as I swim. My eyes blur behind my fogged snorkel mask. I can't see a thing now, but before I can rise to the surface and clean my mask, I sense that I am suddenly surrounded by all six dolphins in the lagoon.

Three ahead of me and three behind me: this is an unusual formation. Usually the dolphins flank me at each side like protective scouts, but now they have placed me within their center in the position of a calf learning to swim. As the three lead dolphins streak forward, their powerful tail flukes pull me in their fast wake. The three dolphins behind me propel me forward with their pectorals, so I need make no movement with my own arms and legs—yet I speed along with them, aerodynamically held in their glide.

Slipstream! My mind recognizes what my body has already enjoyed. The dolphins are swimming for me, carrying me aloft in their waves, just as a dolphin will ride a boat's wake, easily pulled along in another's path. This is how I am honored, as if I am a newborn and can be carried along by the pod.

Never before have the dolphins allowed me this privilege, one usually reserved for their own kind. And in this moment of following, being flung forward without any effort, I feel as if the dolphins understand that they are like rockets, fondly flinging me out into some unknown orbit, some other space that we will no longer share. So fast is this propulsion, it's as if we are no longer in the heavy element of water, but above it. And as I speed along, held in their gravity, an image suddenly comes to my mind. Hundreds of dolphins in the open ocean, porpoising across whitecaps, leaping out of the water together so quickly that their bodies never seem to touch down. Only the tips of their tail flukes splash to show they are swimming, not flying.

"Crossover" is the word scientists use to describe dolphins' soaring over seas, their traveling so free and fast, so high-spirited and almost effervescent that their sleek bodies barely skim the waves. The suggestion of splashes from tail and pectorals leaves a luminous wake across the water. For these crossover miles, the dolphins, like their human terrestrial mammal kin, belong more to the element of air than to sea. Crossover behavior is a wonder to see, and I had not yet witnessed it myself. So where was the vivid image in my mind coming from? The dolphins or me?

As I swam in the dolphins' slipstream, I closed my eyes and watched the bright crossover images expand. Now I saw a superpod of wild dolphins flying over the waves, jubilant and exuberantly free. Had the dolphins carried me not only with their bodies but also with their minds? Was this the hologrammatic intelligence that some dolphin researchers believe exists, what we might experience as a kind of photographic telepathy?

Held in their fluid embrace, I pulled my arms close against my sides and our communal speed increased. The dolphins swam so fast now, swerving at the fences in an expert curve of wake and small waves. Racing around the lagoon, I opened my eyes again to see nothing but an emerald, underwater blur. And then I remembered what I had either forgotten long ago or never quite fully realized. This feeling of being carried along by other animals was familiar.

Animals had carried me all my life. I was a crossover—carried along in the generous and instructive slipstream of other species. And I had always navigated my life with them in mind, going between the human and animal worlds—a crossover myself. By including animals in my life I was always engaging with the Other, imagining the animal mind and life. For almost half a century, my bond with animals had shaped my character and revealed the world to me. At every turning point in my life an animal had mirrored or influenced my fate. Mine was not simply a life with other animals, but a life *because of animals*. It had been this way since my beginning, born on a forest lookout station in the High Sierras, surrounded by millions of acres of wilderness and many more animals than humans. Since infancy, the first faces I imprinted, the first faces I ever really loved, were animal.

And now as these beloved dolphins held me in their moving wake, I followed them one last time. As I glanced over at Alphonse, his brown, benevolent eyes held mine and I tried to send him a mental picture of his own beautiful, wrinkled body at birth. My eyes blurred in my mask as I apologized that I would not be there to see him through his new life. It was a life that broke my heart. Captive-born, he would serve humans and their healing; he would never know a watery world without fences or exult in the company of a wild pod crossing over air and sea.

Straight ahead three dolphins ululated the strong currents that carried me still and as I raced with them, I felt more dolphin than human. In that moment, I knew the crossover images in my mind were not mine alone, but shared. For all the other dolphins carrying me were once free, born in open sea, given a birthright that was now denied. But still they had the memory of speed and slipstream, of spacious ocean and traveling thirty miles a day, their powerful pectorals and tail flukes never flung against fences.

In that lagoon on my last swim with the dolphins whom I will always hold as dear as my own brother and sisters, I promise myself that I will begin the real work of helping to release and restore captive dolphins to the wild. And that perhaps one day before I die, I will hope to be blessed with seeing what these generous ambassador dolphins can only remember: a crossover of thousand-strong dolphins slipping wildly through wide open seas.

The dolphins still remember their wild birthright. And so do I.

The History of the Dolphin

*I have met with a story, which, although authenticated
by undoubted evidence, looks very like a fable.*
PLINY THE YOUNGER

The history of the dolphin is one of the most fascinating and instructive in the historiography and the history of ideas in the western world. Indeed, it provides one of the most illuminating examples of what has probably occurred many times in human culture—a virtually complete loss of knowledge, at least in most segments of the culture, of what was formerly well understood by generations of men. "Not in entire forgetfulness" in some regions of the world, but certainly in "a sleep and a forgetting" in the most sophisticated centers of the western world. . . .

It was Aristotle in his *History of the Animals* (521b) who first classified whales, porpoises, and dolphins at Cetacea, τὰ κήτη οἶου δελφὶς καὶ φωκαὶνα καὶ φάλαινα. Aristotle's account of the Cetacea was astonishingly accurately written, and quite evidently from firsthand knowledge of these animals. . . .

The Common and Bottle-Nosed Dolphins are those best known to the western world, but many of the traits which have recently been rediscovered concerning these creatures have been well known to other peoples for millennia. It is only a certain segment of the western world, its more sophisticated representation, and particularly the learned world, which dismissed as myths the tales told about dolphins in classical antiquity. And this is the real burden of the story I have to tell you. Some of these antique tales may have been myths, but as we shall see, many of them were not, and undoubtedly a number of the myths were based on real events partially embroidered by the imagination and improved, like good wine, by time. But good wine needs no bush, and I shall sample this wine as palatably as I find it.

The earliest representation of a dolphin I have been able to find is from a pictographic seal from Crete, estimated to date from 3500 to 2200 B.C. The earliest *painting* of a dolphin thus far recovered is from the ancient Peloponnesian city of Tiryns. The date is about 1600 B.C. In that city it is also represented in stucco floors. Several good examples of dolphins are furnished by seventh century Corinthian art. The dolphin is also well represented in Minoan art. In Cyprus it is frequently represented in Late Helladic vases, shards, amphorae, in metalwork, engravings, and in stucco floors as at Tiryns. Among the importations from Crete into Helladic art appear to have been certain stylized forms of the dolphin.

An early literary reference to the dolphin occurs in Aesop's fable "The Monkey and the Dolphin." During a violent storm a ship was capsized, and among those thrown into the water was a monkey. Observing its distress a dolphin came to its rescue, and taking the monkey upon its back the dolphin headed for shore. Opposite Piraeus, the harbor of Athens, the dolphin inquired of the monkey whether he was an Athenian. "Oh, yes," replied the monkey, "and from one of the best families." "Then you know Piraeus," said the dolphin. "Very well, indeed," said the monkey, "he is one of my most intimate friends." Whereupon, outraged by so gross a deceit, the dolphin took a deep dive and left the monkey to its fate.

I take it that ever since that day monkeys have very sensibly refrained from speech. It is far better to remain silent even at the risk of being taken for a fool or a rogue, than to open one's mouth and remove all doubt.

Aesop flourished about 600 B.C. His story suggests a considerable knowledge of the ways of dolphins, and this indicates that knowledge of the dolphin was already old in his time.

There are several variant Greek myths on the origin of the dolphin. All of them relate to Dionysos. In one version Dionysos is an adult, in another he is a child. The first group of legends represents the epiphany of Dionysos, symbolizing the battle between winter and summer. Winter is represented by the death of Dionysos who disappears into the water, from which he is brought back on the top of a dolphin as the returning springtime (Apollodorus, III, 5, 3). Another version has Dionysos, whether as child or adult varies, being conveyed by ship to Naxos by Tyrrhenian mariners. The latter conceive the idea of kidnapping him. Dionysos senses their treachery, and bidding his companions strike up on their musical instruments, he produces a Bacchic wild dance in the mariners who throw themselves overboard and are changed into dolphins.

The popular belief in antiquity in the human intelligence of dolphins and their kindly feeling toward man was explained by the ancient writers in the light of the transformation of the Tyrrhenian pirates into dolphins. (See Lucian, *Marine Dialogues*, 8; Oppian, *Halieutica*, I, 649–54, 1098, V, 422, 519f; Porphyry, *De Abstinentia*, III, 16.) As Oppian (I, 1089) in his *Halieutica* has it, in William Diaper's charming translation:

> So *Dolphins* teem, whom subject Fish revere,
> And show the smiling Seas their Infant-Heir.
> All other Kinds, whom Parent-Seas confine,
> *Dolphins* excell; that Race is all divine.
> *Dolphins* were Men (Tradition hands the Tale)
> Laborious Swains bred on the *Tuscan* Vale:
> Transform'd by *Bacchus*, and by *Neptune* lov'd,
> They all the Pleasures of the Deep improv'd.
> When new-made Fish the God's Command obey'd,
> Plung'd in the Waves, and untry'd Fins displayed,
> No further Change relenting *Bacchus* wrought,
> Nor have the *Dolphins* all the Man forgot;
> The conscious Soul retains her former Thought.

The god of the golden trident who rules over the seas, Poseidon, would not have prospered in his wooing of Amphitrite if it had not been for the assistance of a dolphin, who apprized Poseidon of her hiding-place. For this service, as is well-known, Poseidon set the dolphin among the stars in the constellation which bears its name to this day.

It is interesting in this connection that in a modern Greek folktale from Zacynthos, Poseidon changes a hero who has fallen into the sea into a dolphin until such time as he should find a maiden ready to be his wife. After some time the dolphin rescues a shipwrecked king and his daughter, the princess by way of reward takes him for her husband, and the spell is broken (Bernhard Schmidt, *Das Volksleben der Neugriechen*, 135).

The cult of Apollo Delphinus was initiated, so legend has it, by Icadius who, leaving his native land of Lycia, which he had named for his mother, set out for Italy. Shipwrecked on the way, he was taken on the back of a dolphin, which set him down near Mount Parnassus, where he founded a temple to his father Apollo, and called the place Delphi after the dolphin. For this reason the dolphin became among the things most sacred to Apollo (Servius, *Commentarii in Vergilii Aeneidos*, III, 332; also Cornificius Longus, *De Etymis Deorum*).

Herodotus, writing of Periander (fl. 600 B.C.), tyrant of Corinth, tells one of the most famous of all stories of the dolphin (it is mentioned by Shakespeare in the first act of *Twelfth Night*). "In his time," writes Herodotus (b. 484 B.C.),

a very wonderful thing is said to have happened. The Corinthians and the Lesbians agree in their account of the matter. They relate that Arion of Methymna, who, as a player on the lyre, was second to no man living at that time, and who was, so far as we know, the first to invent the dithyrambic measure, to give it its name, and to conduct in it at Corinth, was carried to Taenarum on the back of a dolphin.

He had lived, it is said, at the court of Periander, when a longing came upon him to sail across to Italy and Sicily. Having made rich profits in those parts, he wanted to recross the seas to Corinth. He therefore hired a vessel, the crew of which were Corinthians, thinking that there was no people in whom he could more safely confide; and, going on board, he set sail from Tarentum. The sailors, however, when they reached the open sea, formed a plot to throw him overboard and seize upon his riches. Discovering their design, he fell on his knees, beseeching them to spare his life, and making them welcome to his money. But they refused; and required him either to kill himself outright, if he wished for a grave on the dry land, or without loss of time to leap overboard into the sea. In this strait Arion begged them, since such was their pleasure, to allow him to mount upon the quarter-deck, dressed in his full costume, and there to play and sing, and promising that, as soon as his song was ended, he would destroy himself. Delighted at the prospect of hearing the very best singer in the world, they consented, and withdrew from the stern to the middle of the vessel: while Arion dressed himself in the full costume of his calling, took his lyre, and standing on the quarter-deck, chanted the Orthian [a very high pitched lively and spirited song]. His strain ended, he flung himself, fully attired as he was, headlong into the sea. The Corinthians then sailed on to Corinth. As for Arion, a dolphin, they say, took him upon his back and carried him to Taenarum, where he went ashore, and thence proceeded to Corinth in his musician's dress, and told all that had happened to him. Periander, however, disbelieved the story, and put Arion in ward, to prevent his leaving Corinth, while he watched anxiously for the return of the mariners. On their arrival he summoned them before him and asked them if they could give him any tidings of Arion. They returned for answer that he was alive and in good health in Italy, and that they had left him at Tarentum, where he was doing well. Thereupon Arion appeared before them, just as he was when he jumped from the vessel: the men, astonished and detected in falsehood, could no longer deny their guilt. Such is the account which the Corinthians and Lesbians give; and there is to this day at Taenarum an offering of Arion's at the shrine, which is a small figure in bronze, representing a man seated upon a dolphin. (*The History of Herodotus*, Clio, I, 23–24)

Commenting on this tale the poet Bianor, in *The Greek Anthology* (*Declamatory Epigrams*, 308), remarks, "So the sea presumably contains fish whose righteousness exceeds that of mankind."

Coins of Methymna, in Lesbos, Arion's birthplace, show him riding a dolphin. In one form or another the dolphin is represented on the coins of some forty Greek cities, and doubtless most Greeks knew the reason why.

Pliny the Elder, in his *Natural History* (IX, 8, 24–28), writes as follows:

The dolphin is an animal that is not only friendly to mankind but is also a lover of music, and it can be charmed by singing in harmony, but particularly by the sound of the water-organ. It is not afraid of a human being as something strange to it, but comes to meet vessels at sea and sports and gambols round them, actually trying to race them and passing them even when under full sail. In the reign of the late lamented Augustus a dolphin that had been brought into the Lucrine Lake fell marvellously in love with a certain boy, a poor man's son, who used to go from the Baiae district to school at Pozzuoli, because fairly often the lad when loitering about the place at noon called him to him by the name of Snubnose and coaxed him with bits of the bread he had with him for the journey,—I should be ashamed to tell the story were it not that it has been written about by Maecenas and Fabianus and

Flavius Alfius and many others,—and when the boy called to it at whatever time of day, although it was concealed in hiding, it used to fly to him out of the depth, eat out of his hand, and let him mount on its back, sheathing as it were the prickles of its fin, and used to carry him when mounted right across the bay to Pozzuoli to school, bringing him back in similar manner, for several years, until the boy died of disease, and then it used to keep coming sorrowfully and like a mourner to the customary place, and itself also expired, quite undoubtedly from longing. Another dolphin in recent years at Hippo Diarrhytus on the coast of Africa similarly used to feed out of people's hands and allow itself to be stroked, and play with swimmers and carry them on its back. The Governor of Africa, Flavianus, smeared it all over with perfume, and the novelty of the scent apparently put it to sleep: it floated lifelessly about, holding aloof from human intercourse for some months as if it had been driven away by insult; but afterwards it returned and was an object of wonder as before. The expense caused to their hosts by persons of official position who came to see it forced the people of Hippo to destroy it. Before these occurrences a similar story is told about a boy in the city of Iasus, with whom a dolphin was observed for a long time to be in love, and while eagerly following him to the shore when he was going away it grounded on the sand and expired; Alexander the Great made the boy head of the priesthood of Poseidon at Babylon, interpreting the dolphin's affection as a sign of the deity's favour. Hegesidemus writes that in the same city of Iasus another boy also, named Hermias, while riding across the sea in the same manner lost his life in the waves of a sudden storm, but was brought back to the shore, and the dolphin confessing itself the cause of his death did not return out to sea and expired on dry land. Theophrastus records that exactly the same thing occurred at Naupactos too. Indeed there are unlimited instances: the people of Amphilocus and Taranto tell the same stories about boys and dolphins; and these make it credible that also the skilled harper Arion, when at sea the sailors were getting ready to kill him with the intention of stealing the money he had made, succeeded in coaxing them to let him first play a turn on his harp, and the music attracted a school of dolphins, whereupon he dived into the sea and was taken up by one of them and carried ashore at Cape Matapan.

A very similar but apparently quite independent account of these stories is given by the younger Pliny, in his *Letters* (IX, 23).

The elder Pliny then goes on to tell of the manner in which dolphins assist fishermen, which corresponds closely with the accounts given by recent observers of this cooperative activity between fishermen and dolphins. (For accounts of these see Antony Alpers, *Dolphins*, 146 sq.)

There are numerous other stories similar to those given by the Plinys from classical antiquity, but it is quite impossible to recount them here.[1] What they all have in common is the friendliness of the dolphin for human beings, their rescue of them when they were thrown into the sea, their playfulness, especially with children, and their interest in almost any sort of sound. All these traits came to be regarded as mythical by later and more sophisticated ages, and Usener *(Die Sintfluthsagen)* comments on the effect that the prevalence of these tales had even upon the scientific thought of antiquity, making it difficult for such thinkers as Aristotle to get away from the belief in the dolphin's ability to carry a rider, and in its capacity for human feeling (Aristotle, *History of Animals*, 631a). But Aristotle was right and Herr Usener wrong. The delightful thing about most of these myths is that they all appear to be based on solid fact, and not on the fancies attributed to the original narrators. Another typical modern gloss by a highly sophisticated writer, biologically not unknowledgeable, Norman Douglas, is the following: Commenting on the delphic mythology, he writes,

> From these and many other sources, we may gather that there was supposed to exist an obscure but powerful bond of affection between this animal and humanity, and that it was endowed with a certain kindheartedness and man-loving propensity. This is obviously not the case; the dolphin cares no more about us than cares the haddock. What is the origin of this belief? I conjecture that the beast was credited with these social sentiments out of what may be called poetic reciprocation. Mankind, loving the merry gambols and other endearing characteristics of the dolphin, which has a playful trick of escorting vessels for its own amusement, whose presence signified fair weather, and whose parental attachment to its offspring won their esteem—quite apart from its fabled, perhaps real, love of music or at least of noisy sounds— were pleased to invest it with feelings akin to their own. They were fond of the dolphin; what more natural and becoming than that the

dolphin should be fond of them? (*Birds and Beasts of the Greek Anthology*, 161)

But Douglas was undisillusionedly wrong, and the dolphins are right, and so is the "mankind" that believed in their friendliness. Though pleased to see the dolphins play, it is to be regretted that Douglas did not mind his compass and his way, for:

> Had the curteous Dolphins heard
> One note of his, they would have dar'd
> To quit the waters, to enjoy
> In banishment such melody.
>
> (John Hall, 1646)

In order to avoid any imputation that I may be attempting to play Euhemerus[2] to the dolphin's tale, the facts may be allowed to speak for themselves— always remembering that facts never speak for themselves, but are at the mercy of their interpreters. All, then, that I am concerned to show here, by citing the contemporary evidence, is that, in essence, the so-called myths of the ancients were based on solid facts of observation and not, as has hitherto been supposed, on the imaginings of mythmakers.

Let us begin with a brief account of the most recent and most thoroughly documented story of a free-dwelling dolphin's social interaction with human beings. This is the story of Opo, a female *Tursiops* that made its appearance early in 1955 at Opononi, a small township just outside the mouth of Hokianga Harbour, on the western side of the North Island of New Zealand. From allowing itself at first to be rubbed with an oar or mop carried on the fishermen's launches, it began to glide in near the beach among the bathers. The cheerful *putt-putt* of a motor-launch or of an outboard motor was in irresistible attraction for Opo, and she would follow the boat like a dog, playing or cruising round it. If she had an urge to wander, starting up the motor would invariably draw her back again. Mr. Piwai Toi, a Maori farmer, who was the first to observe Opo, writes, "She was really and truly a children's playmate. Although she played with grownups she was really at her charming best with a crowd of children swimming and wading. I have seen her swimming amongst children almost begging to be petted. She had an uncanny knack of finding out those who were gentle among her young admirers, and keeping away from the rougher elements. If they were all gentle then she would give of her best" (Antony Alpers, *The Dolphin*, 228–29).

The child the dolphin favored was a thirteen-year-old girl named Jill Baker. At fourteen Jill wrote the following account of her experience with Opo:

> I think why the dolphin became so friendly with me was because I was always gentle with her and never rushed at her as so many bathers did. No matter how many went in the water playing with her, as soon as I went in for a swim she would leave all the others and go off side-by-side with me. I remember on one occasion I went for a swim much further up the beach than where she was playing, and I was only in the water a short while when she bobbed up just in front of my face and gave me such a fright. On several other occasions when I was standing in the water with my legs apart she would go between them and pick me up and carry me a short distance before dropping me again. At first she didn't like the feel of my hands and would dart away, but after a while when she realized I would not harm her she would come up to me to be rubbed and patted. She would quite often let me put little children on her back for a moment or two. (In Antony Alpers, *The Dolphin*, 229)

Opo's choice of the gentle Jill Baker for the rides which she gave this thirteen-year-old suggests not only a sensitive discrimination of the qualities of human beings, but also that the reports of similar incidents which have come down to us from antiquity were based on similarly observed events. The one element in these stories which seemed most difficult to accept, and which is so often represented in ancient art, the boy riding on the back of a dolphin, is now removed from the realm of fancy and placed squarely in the realm of fact. It has been corroborated and sustained.

Mr. Antony Alpers in his book on the dolphin, and especially that part devoted to the eyewitness accounts of Opo's behavior, goes far toward establishing the fact of the dolphin's remarkable capacity for rapport with human beings. But for those striking facts I must recommend you to Mr. Alper's charming book.

The dolphin's extraordinary interest in and, what we will I am sure not be far wrong in interpreting as, concern for human beings is dramatically told by George Llano in his report *Airmen Against the Sea*. This report, written on survival at sea during the Second World War, records the experience of six American airmen, shot down over the Pacific, who found themselves in a seven-man raft being pushed by a porpoise toward land. Unfortunately the land was an island held by the Japanese. The friendly porpoise must have been surprised and hurt when he found himself being dissuaded from his pushing by being beaten off with the oars of the airmen.

Dr. Llano also reports that "Most observers noted that when porpoises appeared sharks disappeared, and they frequently refer to the 'welcome' appearance of porpoises, whose company they preferred to that of sharks." This confirms all earlier reports that sharks are no match for the dolphin kind.

Dolphins have been known to push a mattress quite empty of human beings for considerable distances at sea. Possibly it is merely the pushing that interests them, and not the saving of any human beings that might be atop of them.

Is there any evidence that dolphins save drowning swimmers? There is.

In 1945 the wife of a well-known trial attorney residing in Florida was saved from drowning by a dolphin.[3] This woman had stepped into a sea with a strong undertow and was immediately dragged under. Just before losing consciousness, she remembers hoping that someone would push her ashore. "With that, someone gave me a tremendous shove, and I landed on the beach, face down, too exhausted to turn over . . . when I did, no one was near, but in the water almost eighteen feet out a porpoise was leaping around, and a few feet beyond him another large fish was also leaping."

In this case the porpoise was almost certainly a dolphin and the large fish a fishtail shark. A man who had observed the events from the other side of a fence told the rescued woman that this was the second time he had seen a drowning person saved by a "porpoise."

More recently, on the night of February 29, 1960, Mrs. Yvonne M. Bliss of Stuart fell from a boat off the east coast of Grand Bahama Island in the West Indies:[4]

> After floating, swimming, shedding more clothing for what seemed an eternity, I saw a form in the water to the left of me. . . . It touched the side of my hip and, thinking it must be a shark, I moved over to the right to try to get away from it. . . . This change in my position was to my advantage as heretofore I was bucking a cross tide and the waves would wash over my head and I would swallow a great deal of water.

This sea animal which I knew by this time must be a porpoise had guided me so that I was being carried with the tide.

After another eternity and being thankful that my friend was keeping away the sharks and barracuda for which these waters are famous, the porpoise moved back of me and came around to my right side. I moved over to give room to my companion and later knew that had not the porpoise done this, I would have been going downstream to deeper and faster moving waters. The porpoise had guided me to the section where the water was the most shallow.

Shortly I touched what felt like fish netting to my feet. It was seaweed and under that the glorious and most welcome bottom.

As I turned toward shore, stumbling, losing balance, and saying a prayer of thanks, my rescuer took off like a streak on down the channel.

The reader must be left to make what he can of such occurrences. Dr. George G. Goodwin of the American Museum of Natural History doubts the intention of dolphins to save drowning persons.[5] "Anything floating," he writes, "on or near the surface of the sea will attract his attention. His first action on approaching the object of his curiosity is to roll under it. In doing so, something partly submerged, like the body of a drowning person, is nudged to the surface of the water. The sea does its part and automatically drives floating objects toward the beach." This may well be so in some cases, but it is an explanation which does not fit the incidents described by Mrs. Bliss, in which she was not pushed but guided. Occam's razor should not be too bluntly applied.

The cooperativeness of dolphins with fishermen in various parts of the world has gone on for several thousand years without its significance having registered much upon the consciousness of the rest of the world—including the learned and the scientific.

In the Mediterranean from the earliest days, as recorded by Aelian in his *On the Characteristics of Animals* (VI, 15) to the present day, torchlight fishing with the aid of dolphins has been a traditional way of fishing. This has been described by Nicholas Apostolides in his book *La Pêche en Grèece*, who tells how fishermen of the Sporades catch their garfish "in the darkest nights of the month of October" by methods very similar to those described by Aelian. Briefly, the fish attracted by the fishermen's flares begin to collect, whereupon the dolphins appear and drive them into the fishermen's nets.

Similar methods of fishing were practiced in the Antipodes, off the New Zealand and Queensland coasts. The aborigines of Moreton Bay, Queensland, used to catch mullet with the aid of dolphins, at a place appropriately enough called Amity Point. The aborigines recognized individual dolphins and called them by name. With their nets ready on the beach the aborigines waited for a shoal of fish to appear, whereupon they would run down and make a peculiar splashing in the water with their spears, and the dolphins on the outside of the shoal would drive the fish towards the nets for the aborigines to catch. Fairholme, who described these events in 1856, writes, "For my part I cannot doubt that the understanding is real, and that the natives know these porpoises [actually the dolphin *Tursiops catalania*], and that strange porpoises would not show so little fear of the natives. The oldest men of the tribe say that the same kind of fishing has always been carried on as long as they can remember. Porpoises abound in the bay, but in no other part do the natives fish with their assistance."

The Irrawaddy River dolphin is also an assistant-fisherman. John Anderson reports that "The fishermen believe that the dolphin purposely draws fish to their nets, and each fishing village has its particular guardian dolphin which receives a name common to all the fellows of his school; and it is this superstition which makes it so difficult to obtain specimens of this Cetacean. Colonel Sladen has told me that suits are not infrequently brought into the native courts to recover a share in the capture of fish, in which a plaintiff's dolphin has been held to fill the nets of rival fishermen" (John Anderson, *Account of the Zoological Results of Two Expeditions to Western Yunnan*).

The pink-bellied river dolphin *(Inia geoffrensis)* of the Trapajós, a tributary of the Amazon, also helps its human friends with fishing. Dr. F. Bruce Lamb[6] says that this dolphin, locally known as the *boto*, "is reported to have saved the lived of helpless persons whose boats have capsized, by pushing them ashore. None of the dreaded flesh-eating *piranhas* appear when a porpoise is present, for they themselves would be eaten." And he goes on to give an eye-witness account of fishing with the aid of a trained dolphin. "My curiosity was aroused," he writes, "by the paddler, who began tapping on the side of the canoe with his paddle between strokes and whistling a peculiar call. Asking Rymundo about this, he startled me by casually remarking that they were calling their *boto*, their porpoise. . . . As we approached the fishing grounds near the riverbank, Rymundo lit his carbide miner's lamp, adjusted the reflector, chose his first harpoon, and stood up in the bow ready for action. Almost immediately on the offshore side of the canoe about 50 feet from us we heard a porpoise come up to blow and take in fresh air." The porpoise then chased the fish toward the canoe and Rymundo harpooned them with ease.

Many ancient writers have referred to the brilliancy of the changeful colors when the dolphin is dying. Byron makes reference to this in "Childe Harold's Pilgrimage,"

> Parting day
> Dies like the dolphin, whom each pang imbues
> With a new colour as it gasps away;
> The last still loveliest, till 'tis gone, and
> all is gray.

Here is a peculiar confusion, for this is not the mammalian dolphin of which we have been speaking but the swift piscivorous oceanic fish *Coryphaena hippurus*, the dolphin of sailors. It is blue with deeper spots, and gleaming with gold. It is, indeed, famous for the beauty of its changing colors when dying. The mammalian dolphin exhibits no such spectacular color changes when dying.

Happily, it is not with dying dolphins or with *their* changing colors that we are concerned here, but rather with ours, the changing color of the complexion of our once too sophisticated beliefs. Beliefs which, in their own way, were very much more in the nature of myths than the ancient ones which we wrote off a little too disdainfully as such. The history of the dolphin constitutes an illuminating example of the eclipse of knowledge once possessed by the learned, but which was virtually completely relegated to the outermost fringes of mythology during the last eighteen hundred years. Perhaps there is a moral to be drawn here. If so, I shall leave it to others to draw. But now that scientific interest in the dolphin has been aroused we are entering into a new era of delphinology, and with the confirmation of so many of the observations of the ancients already made, we may look forward with confidence to others. Dolphins have large brains; possibly they will some day be able to teach us what brains are really for.

NOTES

1. Among the many well-known figures of classical mythology said to have been saved by dolphins from the sea are Eikadios, Enalos, Koiranos, Phalanthos, Taras, etc. In many other cases the corpses were brought ashore by a dolphin, which then expired on reaching land (similarly, with minor variations, was this so with Palaimon, or Melikertes, Dionysios and Hermias of Iasos, Hesiod, and the boys already referred to from Baiae and Naupaktos). Similar incidents reappear in the writings of the hagiographers. Saints Martinianos of Kaisareia, Kallistratos of Carthage, and Basileios the younger of Constantinople were each saved from a watery grave by a couple of dolphins. The corpse of Saint Loukianos of Antioch was brought ashore by a large dolphin, which then expired on the sand. See Klement, *Arion*, 1–64, and Usener, *Die Sintfluthsagen*, 138–80.

2. Euhemerus (ca. second half of the fourth century B.C.) attempted a rationalistic explanation of the mythology prevailing in his time. The theory he propounded, in his novel of travel, *Sacred History*, was simply an extension of the current skeptical-scientific attitude to matters which until that time had been accepted without question. That theory was that the gods were merely men who because of their great exploits or beneficence had been accorded divine honors. In Crete, coming upon the remains of a tomb bearing the name of Zeus, Euhemerus argued that even Zeus had probably been no more than a great conqueror, who died and was buried in Crete, and afterward deified. This creditable anthropological attempt to historicize mythology, though it failed to convince, is nevertheless worthy of great respect. As A. B. Cook wrote, if Euhemerus said that Zeus was a Cretan king when he ought to have said that Cretan kings played the part of Zeus, it is a pardonable error (*Zeus*, I, 662).

3. "Saved by a Porpoise," *Natural History* 58 (1949): 385–86.

4. Winthrop N. Kellogg, *Porpoises and Sonar* (Chicago: University of Chicago Press, 1962), 14.

5. George G. Goodwin, "Porpoise—Friend of Man?" *Natural History* 56 (1947): 337.

6. F. Bruce Lamb, "The Fisherman's Porpoise," *Natural History* 63 (1954): 231–32.

At-One-Ment

Humans love playing with other animals, and sometimes this leads to a purity of exchange almost magical in its intensity, deep play at its best. For instance, I once heard about a friendly group of spotted dolphins that are drawn to music played underwater and readily swim with divers in the warm currents of the Bahamas. Research teams visit yearly to chronicle their history and habits. Villagers in the sea, the dolphins form a community that changes as couples mate, young are born, the aged die, and new alliances are forged. There is nothing like the indelible thrill of meeting a wild animal on its own terms in its own element, so I decided to join a week-long trip. One morning I flew to Grand Bahama, took a cab to the West End, and boarded a two-masted schooner along with eight other researchers. We were hoping to encounter spotted dolphins often enough during the week at sea to be able to identify and catalogue individuals.

At six-thirty the following morning, we left the West End behind and cruised toward the Little Bahama Bank, a shallow area that spotted dolphins seem to prefer. After a few minutes, we hoisted the mainsail, from which two rows of short ties hung like fringe. Then we sat on benches or low deck chairs, finding shade under a large blue canopy stretched over the center of the boat and attached by a web of ropes over the boom. The ocean poured blue black all around us with rose-gold shimmers from the sun. Gradually, the water mellowed to navy blue, then indigo, and finally azure, as we drew closer to the shallows. Clumps of turtle grass looked like cloud shadows on the floor. After three hours, a pale-blue ribbon appeared on the horizon and we headed toward it. Flying fish leaped near the bow and hurled themselves through the air a dozen yards at a time, like rocks skipping over the water.

Soon we entered the dreamtime of the aqua shallows. This area rises like a stage or platform in the ocean, without coral or large schools of fish. It appears to be a desert, a barren pan; but there are few places on earth without life of some sort. Here there is a bustling plant community, from simple blue-green algae to more complicated plants with stems and leaves. Plankton, the first step of the food chain, thrives on the banks, even though the waters look quite empty to the casual observer.

A bottlenosed dolphin leaped near the boat, then zoomed in and lined up with the bow, swinging back and forth like a surfer finding the sweetest spot of a wave. Soon it was joined by a second. Hobos hitching a ride, the dolphins weren't moving their tails at all, but were carried along at speed by the bow wave. They seemed to relish the sport. Beautiful as these dolphins were, we were on the lookout for their cousins, the spotted ones.

A Concorde sailed through the ocean overhead, making a double *boom!* as it passed. What is speed to the passengers on that supersonic, I wondered, or to the dolphins surfing on the bow wave?

"More dolphins!" the captain cried, pointing west.

As seven spotted dolphins homed in on the boat, we donned snorkeling gear and jumped into the water with them. A mother and baby accompanied by another female arrived first, swam straight up to us and started playing. A dolphin went close to one woman, waited for her to follow, then started turning tight circles with her. Like dervishes, dolphin and human spun together. Meanwhile, two other dolphins dived down to the bottom, about forty feet below, and made fast passes at me. I turned to follow them. Slowing, they allowed me to swim with them in formation, only inches away. By now the dolphins were all over us, swirling and diving, coasting close and wiggling away to see if we'd follow. If I dived, they dived, and they often accompanied me back to the surface, eye to eye. At first it was startling how close they came. We have an invisible no-man's-land around our bodies that others don't enter unless they mean to romance or harm us. To have a wild animal enter that dangerous realm, knowing that you could hurt it or it could hurt you, but that neither of you will, produces instantaneous trust. After it happens once, all fear vanishes. Somehow, they managed to keep their slender distance— as little as two or three inches—without actually making physical contact.

But they were touching in another sense, with their X-ray-like sonar, patting our skin, reaching deep inside us to our bones and soft tissues. At times, I could feel their streaming clicks. They seemed especially interested in one woman's belly. Could they tell she was pregnant? Probably. I wonder how the fetus showed up on their sonar. Could they echolocate our stomachs and know what we ate for lunch? Could they detect broken bones and tumors? Could they diagnose some diseases in us and in themselves? Hard to say. Because we couldn't touch them, they seemed aloof. But they were touching us constantly. For them, the contact was intimate, sensuous, if one-sided. What do dolphins feel when they echolocate one another?

At least we know *how* dolphins echolocate: they produce narrow streams of clicks (intermittent bursts of sound that last less than a thousandth of a second each) by blowing air back and forth through nasal passages. When the sound enters a fat-filled cavity in the head, it's focused into a single beam that can be directed wherever the dolphin wishes. First the dolphin sends out a general click, then it refines the signal to identify the object, which usually takes about six clicks, each one adjusting the picture so subtly that only chaos theory can explain it. At lightning speed, the dolphin sends out a signal, waits for the echo, decides what pulse to send next, waits for that one's echo, and so on, until it detects the object and classifies it. Some likely categories are: edible, dangerous, sexy, inanimate, useful, human, never-before-encountered, none of the above.

For an hour, the dolphins played exhausting, puppyish chase-and-tumble games. Meanwhile, we tried to study their markings. Each had a distinctive pattern of spots, tail notches, blazes. Often they darted to the sandy bottom and found silvery sand dabs that they chased and ate. They were like hyperactive children, easily bored, full of swerve and spunk. And we were their big bathtub toys. A pink ribbon floated loose from my long braid. In a flash, the dolphin grabbed the ribbon, then tossed it up, caught it with a flipper, tossed it backward, kicked it with its tail, caught it with the other flipper, spun around, slid it over its nose, swam away with it, then returned a moment later and let it fall through the water like a cast-off toy. A clear invitation. Taking a lungful of air, I dived after the ribbon, grabbed it with one hand, tossed it back to my fins and flicked it with a clumsy kick. By this time I needed to resurface to breathe, so I let the ribbon drift down, undulating like a piece of kelp. The dolphin collected it at speed over the pyramid of one flipper, let it slide back to the tail, whisked it up and tossed it with the other flipper. I knew the rules of the game, but I didn't have the breath to play it. Even if I were a pearl diver and could hold my breath for over five minutes, I would still have been out of my league. After a few more of my clumsy lunges for the ribbon, the dolphin swam away, turning its attention to a human who could stay underwater longer, a man taking still photographs with a flash camera.

When at last they veered off toward the horizon, we gathered under the blue canopy to fill in sketch sheets and record the details of each animal. These rough sketches would be compared to photos of known animals, and become part of the researchers' catalogue. You'd think nine observers might supply the same facts, but we didn't all agree on what we saw. Indeed, we sounded like people comparing different versions of an accident. As I filled out sketch sheets, I scanned my memory for head, tail, and flank markings. Living mainly in the tropics, spotted dolphins have long narrow beaks and can be heavily freckled. Like reverse fawns, the young begin life solid colored, usually gray, and only develop white spots as they age. By the time they're elderly, they're covered in swirls of spots and splotches. They grow to about eight feet long, have teeth, and are gregarious. They love to play, which they do with endless ingenuity and zest. Athletic, acrobatic swimmers, they leap into aerial pirouettes, cartwheels, and what seem like attempts to see how long they can hover in the air.

Over the next week, we encountered spotted dolphins every day, the longest session lasting three hours, so long in fact that we were the first to give up out of exhaustion, only to find the dolphins racing after us and trying to tempt us back to romp. Mother dolphins often brought the sleek little surprise of their babies, which appeared perfect and unmarked by life. Sometimes a baby would swim tucked underneath its mother, making a crescent shape, so that it looked as if the baby were still being carried inside. We grew to know them as individuals, a rare privilege. In our travels across the banks, we played with dolphins nine times. The most frequent visitor was a particularly rambunctious five-year-old female called "Nicky," and she became a special favorite. Often, the dolphins arrived like a visitation. Long hours of waiting, in a slow-motion of heat, glare, and water, were suddenly broken by the wild and delicious turmoil of incoming dolphins. When they left, everything fell calm again and we waited once more, at a low ebb, under the harrowing sun.

On our last day, after a particularly exhausting afternoon with Nicky and her friends, we gathered on the deck to watch the sunset. These were some of the most dramatic moments in each day, when the soft aqua of the water fanned through rainbow blues and was washed away in the molten lava of the setting sun. Night fell heavily, in thick black drapes. Retreating to the galley downstairs, we sat and talked about the week. Despite the mild discomforts of ocean sailing, everyone was sad the voyage was over, and all felt nourished by a week of such intimate play with wild animals. I was especially surprised by how eager the dolphins were to make eye contact. Their wildness disappears on one level and is enhanced on another when you stare straight into their eyes, realizing that these are wild creatures and there is something special happening inside their minds. At the very least, there is a willing gentleness and an awareness that draws you in. One reason the plight of the dolphin touches us is because we fear they may be self-aware, not just meaty animals but intelligent life-forms. Suppose, like us, they have inner universes? Suppose they are not like elk or salmon, but animals with a culture of sorts, animals that can judge us?

During the night, the winds kicked up and four-foot seas rolled in from the southwest. Sleeping on deck, I awakened to find that I had slid off my air mattress and my legs were suspended over the side of the boat. Hands folded on my chest, I looked set for burial at sea. So I retreated below, wedged myself into a narrow bunk, and tried to sleep, which was nearly impossible given the lurching and shuddering of the boat. My thoughts turned frequently to the dolphins. Where were they now? What were they doing in the incomprehensible darkness of sea and greater darkness of night? I was stricken both by our kinship with them and by the huge rift between us evolution has created. They were minds in the ocean long before we were minds on the land. They abide by rhythms older than we know or can invent. We pretend we can outsmart them and ignore such rhythms, but in our hearts we know we're steered by them.

In all the excitement of the week, people had leaped into open ocean and swum with schools of large bullet-shaped barracudas, seen a bull shark at touchable range, watched lavender skates hiding on the sandy floor beneath them, and found frisky, hospitable dolphins everywhere. That made the ocean itself seem friendlier. Lumbering creatures of the earth, we find the ocean frightening. For most people, it is another form of night. It seems dark, endless, hostile, full of monsters ready to separate us from our cells. Dolphins leap from that world with a Mona Lisa smile. Playing recognizable games, they reassure us that the unknown may not be so hostile after all.

Throughout history, humans have been enthralled by the dream of swimming and playing with dolphins. The urge is ancient and powerful, and often appears in religious myths. It possesses us, but why? Perhaps, in part, because it seems so much like flight. Who hasn't felt the euphoria of flying in dreams? There is a peace in weightlessness, a freedom that comes from breaking the physical bonds that hold us. We are beings inextricably anchored to the ground, who walk the earth and, in time, will become part of it. In rare moments, or in dreams, we rise above those grim restraints and joyously take flight. Perhaps, also, on some level, flying recalls for us the perfect peace the embryo feels afloat in its mother's womb. All this echoes the weightlessness we feel in the sea, a victory over gravity symbolized by the rapture of dolphins. We want to regain that buoyancy for ourselves. Then, unfettered, our minds might roam more freely. "To halt and hang attached to nothing, no lines or air pipe to the surface, was a dream," Jacques Cousteau writes of his ocean dives. "At night I often had visions of flying by extending my arms as wings. Now I flew without wings." We know dolphins are intelligent, perhaps as intelligent as we humans. In Shark Bay in Western Australia, for example, dolphins have been seen carrying sponges on the tips of their snouts. Researchers think the dolphins are using the sponges to protect their noses from spines and stings as they hunt food on the seabed, and dragging sponges in the sand to flush out potential prey. And there is that fixed smile. Although it's an anatomical accident, not a voluntary expression, it makes them look as if they might be having more fun than we stodgy folk above the waves. Small wonder people credit them with brainpower, spirituality, and deep emotions. Aristotle even claimed that dolphins speak like humans, and "can pronounce vowels . . . but have trouble with consonants."

We sometimes picture dolphins as cherubic creatures, or benign space aliens, a New Age emblem of otherworldly innocence. But they're also seen as powerful shamanistic envoys. Dazzling as they leap from the thick churning water into the sheer invisibility of air, they seem immune to boundaries, magical in flight. They transcend natural prisons. They bridge worlds. Dolphins are often portrayed as seers and savants who can peer into our hidden depths, those dark emotional oceans inside our psyches. To some, they dwell in a world of crystals and spirit guides. To others they allow us a transmigration of souls. In cultures as varied as Greek, Inuit, and that of the Australian Aborigines, art and myth show dolphins saving people, guiding ships to safety, playing with swimmers, having a special kinship with humans. In Belém, on the Amazon, one can find river dolphin vulvas and penises for sale in the voodoo market. But on most of the Amazon the pink dolphins are protected by elaborate myths and no one hunts them. Like the silkies of Scottish legend, dolphins are thought to come ashore and make love with women from time to time. So an illegitimate child is said to be the child of a dolphin, and if you kill a dolphin you may be killing your own father. In Native American myths, dolphins are messengers from the Great Spirit. In Arabian tales dolphins accompany the souls of the dead to the underworld. To early Christians, a dolphin draped over a cross symbolized Christ. The Minoans revered the dolphin, proclaiming it an incarnation of their sea god, Poseidon. Born of Apollo (the sun) and Aphrodite (love), it connected the blazing sky with the fertile sea. In Greek, the words "dolphin" (*delphis*) and "womb" (*delphys*) sound alike. Say *dolphin*, and be reminded of the womb of the sea; say *womb* and be reminded of the spirit of the dolphin connecting male and female, heaven and earth. Greek and Roman mythology abounds with gods and goddesses, sailors and other mortals, being transformed into dolphins. Thanks to Apollo, a dolphin even swims across the night sky as the constellation Delphinus. Ovid tells the story of Arion, a seventh-century B.C. poet and musician, born on the isle of Lesbos, who traveled throughout the kingdom, singing and playing the lyre. Returning home from Sicily with his earnings, he was robbed by sailors who intended to kill him. Arion asked to sing one last song before he was murdered, and permission was granted. His poignant music traveled across the waves, attracting a school of dolphins who encircled the boat. Arion leapt overboard into their midst and a dolphin carried him on its back to shore. There he reported the thugs, who were captured and punished. A great lover of poetry and music, Apollo was so pleased with Arion's triumph that he placed the dolphin and the lyre among the constellations.

Dolphins, running before the bow of a ship, connect the worlds of water and air so deftly that they seem at times to be not only spirit guides, but ambassadors who might somehow intercede between us and the rest of nature. Perhaps there is a need in us to feel accepted by dolphins on behalf of nature, a need for a spiritual experience, a need to reassure them out of a sense of collective guilt that we mean them well. Finding a way back into an intimate relationship with nature may well be the essence of our age-old dolphin dreams. But what are we to make of dolphins and humans *playing* together, seemingly with equal abandon? Dolphins choose to play with us when they find us in their waters, but we travel long distances and endure considerable hardship to make that communion possible, to enjoy deep play with wild animals. On returning from such pilgrimages, people speak like mystics of the transcendence they felt and the transforming beauty of the ocean, which they describe as a sacred realm.

Being with Dolphins

Humans seem to be magically drawn to dolphins. These aquatic mammals are sleek and smooth and glide along apparently without effort. They live in close-knit societies, and mothers take exquisite care of their young. They are always caressing each other: a flipper touch, a close approach, a hint of sexual interaction.[1]

We know by now that dolphins are more than simply idealized mirrors of our own species. Like us, they are full of political intrigue, apparent jealousies, and aggressive ways.[2] As social mammals, they cannot be otherwise. They coordinate their lives superbly well, but it is also a tough life out there: searching for and securing suitable mates, finding enough food to survive, keeping baby away from large sharks and aggressive males, riding out a hurricane or typhoon, dying—and keeping from dying—during near-shore red tides of dinoflagellate blooms, and coping with immense parasite loads and the ever-increasing threat of human pollution.[3]

Despite this knowledge of dolphins as complex animals, we still want to go near them, interact with them, watch and learn and dream and admire. We approach them by boat and they leap toward us to ride the bow wave, cavorting in it and showing joy. We swim with them and watch them underwater as they pirouette and dance around us, curious about this other intelligent stranger in their realm; we are convinced that they are enjoying us as we enjoy them, a blend of species and minds. Surely we cannot be doing them harm by being there, just as they are not harming us.

Let us consider this interaction more deeply. Dolphins do have a tough life out there, and their lives are not divided into segments of work and leisure as ours are. We come to them on our vacations, on our terms. When we interact with them, when they ride the bow wave or cavort with us underwater, it is almost always only a small subset of the animals present. In a group of a hundred Hawaiian spinner dolphins resting in daytime in a near-shore bay, perhaps only ten actually "enjoy" the presence of the humans. And even these ten are taking time away from rest, meaningful social interactions among their own kind, or taking care of young.[4] This smaller proportion of the main group "enjoys" us, even if only briefly, and reinforces the belief that "all the dolphins like us."

After years of being saturated with media images, tourists expect a performance from dolphins. This is the downside of encountering dolphins through animal shows, where their behavior is edited for human consumption and the human demand to be entertained. It may well be that our coming to them at a quiet time is actually a gross infraction of their space: we awaken them to tell them we love them, and they—while some do appear to at least briefly enjoy the interaction—are spending energy with us that might be spent on themselves. Would it not be better to wait for them at a discreet distance—say 300 yards—and let *them* come to us if they so choose?

As researchers, we have worked with dolphins for about thirty years now, in open-water, near-shore, and river environments. We know that dolphins have distinct behavior patterns or "moods," as do mammals in general. When dolphins are feeding or sexually busy, they tend to ignore us and other outside distractions. This is a good time for researchers to view them; we can do so with the reasonable assumption that we are not bothering them much or at all. However, this quiet phase of their behavior may also make them relatively "boring" to the tourist who wants them to be up close, cavorting around the bow, leaping, or swimming tight curious circles around the humans in water. The dolphins are busy and, thankfully, are not easily disturbed. The danger is that the persistent tourist (or researcher) who comes too close or approaches too rapidly may snap the dolphins out of their self-involvement. They stop feeding, or change interactions, and we have committed a disturbing act. Unfortunately, unscrupulous or ignorant tour operators or swim leaders will purposefully approach self-involved animals in an aggressive manner (but they never, ever, admit to this), because it gives them and their charges "a better show."

When dolphins are in deep rest, they tend to be very difficult to approach by boat or underwater. They actively avoid the approaching humans, even doubling under or around the disturbance, or diving for inordinately long times. As researchers, we know this behavior particularly well, because it is possible to totally lose sight of a group of fifty resting dolphins on a flat calm sea. Interestingly, this same evasive behavior is manifested by dolphins who have not fed at all, or not fed much, for one to several days. They are looking for food, they are probably hungry, perhaps even with a low blood sugar concentration as happens in humans, and any outside stimulus is likely to be taken as a disturbance and is therefore cause for evasive action. People who persist in approaching the dolphins when they are evasive will definitely disturb them. The dolphins spend considerable energy in their evasive behavior and presumably forgo true rest or efficient foraging while doing so. Unfortunately, this type of disturbance is often the case, especially with dolphins that rest during daytime and feed in deep waters at night, masked to the human by "that 10 percent" of animals who allow themselves to be waylaid into curiously approaching the humans. Be assured that they do not like being with you when they are engaged in evasive behavior, and that almost all are working hard to get away from you.

To be sure, there are times when dolphins are in a highly social and quite apparent "happy" mood. It is probable that they have fed and rested well, and their party atmosphere—with much aerial displaying at the surface, rapid interactions underwater, and a strong tone of sexual interaction—bleeds over from intraspecific to interspecific behavior. Dolphins in this mood will sometimes play with sea turtles, marine birds, pinnipeds, kelp strands, and humans in the water. There may not be a specific direct bias toward play with humans, but the generalized outgrowth of dolphin socializing sometimes includes humans as well. Their overall mood seems so carefree and happy that we human observers might convince ourselves that we cannot possibly be harming them by approaching them at this time. And, we must always ask: what do the dolphins need, quite separate from us? Their primary behavioral goal is socializing, which probably evolved as an important mechanism for strengthening bonds. We agree, with one significant caveat: it is difficult even for experienced researchers to ascertain from the surface whether it is indeed all or most of the dolphins who are in this carefree-seeming mood, or whether—once again—it is "that 10 percent" fringe.[5]

Researchers who study the behavioral ecology of dolphins do so with a variety of tools: theodolite[6] or otherwise remote-sensing from shores and airplanes, recording vocalizations underwater, and describing group configurations and surface activities from a distance.[7] But much of our observation relies on close approaches to photo-identify and describe interanimal focal behavior from only yards away. We slowly approach, keeping our engine at low speed, avoiding rapid shifts in power and direction. If we dive with dolphins, we do so without disruptive scuba and keep motions as smooth as possible. We do not rapidly extend arms or kick legs, because such actions (often resembling "flailing," common for novice swimmers/divers) are certain to startle.

Dolphins are smart animals, though, and they learn very rapidly whether a novel stimulus means harm or not. In addition, the very novelty of our being near them wears off for them after a while, and we appear to be accepted into their watery world, or perhaps merely tolerated. Our "gentle" insinuation into their society may therefore be similar to research on chimpanzees and other great apes, begun in earnest by Jane Goodall over forty years ago.[8]

Why, then, worry about tourism? Well, in general, tours tend to come and go much more rapidly than do researchers, and habituation is not as likely to occur in a short time frame. Also, all too often boat-based tour operators wish to "give their clients a good show" by driving particularly close or fast in order to elicit reactions: bow riding, tail slapping, or leaping. However, when tourists follow common-sense rules and practice a bit of patience and restraint, they too can be accepted, especially when the objects of their affections are not resting or searching for food.

Although we are content that habituation can take place more readily by our careful approach, it is often surprisingly difficult to gauge the disturbance reactions of animals from the disturbing vessel itself. This includes swimmers in the water. Perhaps the dolphins already changed their behavior in response to us before we were able to observe them ourselves. Perhaps most of the dolphins moved away, but it is the tolerant ones who stayed. Perhaps all animals are being affected, but on a subtle behavioral or physiological level of increased nervousness, adrenalin flow, heart rate, or stomach peristalsis.

A terrestrial analogy might be appropriate here. If a car stops by the side of the road where a group of deer is grazing, the deer may all take fright and leap off. We know that we disturbed them, just as we know that we are disturbing dolphins from their rest when we approach them aggressively. However, if there is no obvious reaction as the car stops, we still do not know whether or not the aforementioned physiological reactions are taking place. Detailed studies on deer and other terrestrial animals have shown increased vigilance behavior, decreased feeding, and a general (much of the time very slow) edging away from the foreign object. Such subtle reactions can have long-term implications on health, fecundity, and longevity.[9] By the way, if the car does not stop but merely passes by at regular speed, the deer would probably (near a reasonably active road or highway) not react at all, behaviorally or physiologically. They have "tuned out" this nonthreatening part of their perceptual world. Similarly, dolphins (and whales) seem to tune out the movements of fishing vessels, predictable ferries, barges, and other industrial traffic in busy shipping lanes. It is the tourist or researcher who orients toward and stays with the animals who is likely to elicit at least that initial reaction or change in behavior. In the dolphin "passing by" case, of course, we still cannot rule out physiological reactions, simply because we have not yet studied them.

There are several indications of long-term effects of human interactions on dolphins in the two areas where we have studied them most intensively. In Kealakekua Bay, Hawaii, spinner dolphins are now residing for shorter times per day and about 25 percent fewer days in the bay than before swimmers and kayakers started invading their daytime rest in the early 1980s.[10] Off Kaikoura, New Zealand, dusky dolphins now travel much farther south in summer months, up to thirty miles away from both organized tourism and dozens of private boats, than they used to before any tourism—again—in the early to mid-1980s. In both cases of overall drastic and potentially important long-term changes in habitat use, there is no direct "proof" that tourism was (and is) responsible. The shifts could be due to a change in ecology, perhaps changes in nighttime food distribution along the respective islands. We do not know and, for these two cases, are unlikely to ever know.

In an uncertain world, a reasonable measure of the Precautionary Principle should be applied. It essentially advises that environmental managers should assume human-caused problems if these are indicated but unproved.[11] It does not help to demand "one more study" (and another one after that) in order to prove disturbance, or other environmental insult, before calling people to action. In the present case, we would advise that human involvement with dolphins in Kealakekua Bay be strictly regulated and that the dolphins be given their usual rest area as a sanctuary where no vessels or human swimmers would be allowed to go. Unfortunately, this also means human enforcement, and fines for scofflaws. The regulations of fines and procedures are already in place in the United States, but the will of the designated enforcers is weak; nothing has happened for over ten years of unregulated human insult to the dolphins. In New Zealand, the situation is much better, because intelligent tour operators have seen the need for caution themselves and have instituted a self-enforced "time out" period of no dolphin approaches for several hours during midday. In addition, the more environmentally conscious people uphold the government's argument that no more than the currently allowed number of boats should be near the dolphins, with strict enforcement of vessel and swimmer behavior near them. Nevertheless, more could be done, and we would like to know whether the greater movements of dolphins away from their previous habitats can be ameliorated by further human action.

We cringe slightly when we hear of another area that has opened to tourism and "swim-with-the-dolphins" programs. We would prefer that the animals—especially in those near-shore areas where they often rest, take care of their newborn, feed in specialized ways, or avoid open-ocean predation—be admired from afar. We personally prefer to study them from cliffs or other promontories, where we know the animals are not disturbed by us and we can see "true behavior." But this is not likely to happen in tourism aimed at dolphins, given our human propensity for wanting to be as close as possible to the action.[12] With this reality, we are relatively content to help push for protective legislation and for education. If you cannot stay away from dolphins, please keep a respectable distance. Insist that you, your guides, and your fellow tourists respect the dolphins' space, allow them to come to you if they so desire, maneuver gently around them. Above all assume—always assume—that they can be disturbed by you even with that "smile" frozen on their faces by the forces of evolution. Dolphins are indeed highly intelligent members of life on earth, and they deserve more true respect, and perhaps less romanticizing of their marvelously well-adapted ways, than we often give them.

NOTES

1. D. A. Pabst, S. A. Rommel, and W. A. McLellan, "The Functional Morphology of Marine Mammals," in *Biology of Marine Mammals*, ed. J. E. Reynolds III and S. A. Rommel (Washington, D.C.: Smithsonian Institution Press, 1999).

2. R. C. Connor, A. J. Read, and R. Wrangham, "Male Reproductive Strategies and Social Bonds," in *Cetacean Societies: Field Studies of Dolphins and Whales*, ed. J. Mann, R. C. Connor, P. L. Tyack, and H. Whitehead (Chicago: University of Chicago Press, 2000).

3. J. R. Twiss, Jr., and R. R. Reeves, eds., *Conservation and Management of Marine Mammals* (Washington, D.C.: Smithsonian Institution Press, 1999).

4. K. S. Norris, B. Würsig, R. Wells, and M. Würsig, *The Hawaiian Spinner Dolphin* (Berkeley: University of California Press, 1994).

5. B. Würsig, M. Würsig, and F. Cipriano, "Dolphins in Different Worlds," *Oceanus* 32, no. 1 (1989): 71–75.

6. The theodolite is a surveyor's transit that allows quite accurate plotting of dolphins and whales on the surface of the water as well as any potential sources of disturbance, such as a boat. If one knows the height of one's station on shore, one can use the surveyor's transit to obtain horizontal and vertical degree readings, and therefore positions, speeds, orientations, etc., of animals.

7. A. Samuels and P. Tyack, "Flukeprints: A History of Studying Cetacean Societies," in Mann et al., eds., *Cetacean Societies*.

8. J. Goodall, *The Chimpanzees of Gombe: Patterns of Behavior* (Cambridge, Mass.: Harvard University Press, 1986).

9. W. J. Richardson and B. Würsig, "Significance of Responses and Noise Impacts," in *Marine Mammals and Noise*, ed. W. J. Richardson, C. R. Greene, Jr., C. I. Malme, and D. H. Thomson (San Diego, Calif.: Academic Press, 1995).

10. A. Forest, "Spinner Dolphins of Kealakekua Bay: Effects of Tourism" (Master's thesis, Department of Wildlife and Fisheries Sciences, Texas A&M University, 2000).

11. Food and Agricultural Organization of the United Nations, *Code of Conduct for Responsible Fisheries* (Rome: FAO-IUCN Press, 1995).

12. While we decry unregulated or poorly regulated whale and dolphin tourism in the developed world, tourism is obviously more desirable than killing the animals. In protein-poor areas of Earth, dolphins often are known only from the local fish market. How much better it is for fishermen to take tourists to see dolphins than to go out and spear them to sell! Not only does the former potentially provide a new livelihood to a waning fishing industry, but it can also show the local people what magnificent animals they have off their beaches and begin the process of awareness of this aspect of nature. This argument is expounded by W. F. Perrin, "Selected Examples of Small Cetaceans at Risk," in Twiss and Reeves, eds., *Conservation and Management of Marine Mammals*.

The Kindred Wild

*Kinship is universal. The orders, families, species and races
of the animal kingdom are the branches of a gigantic arbour. . . .
Man is simply one portion of the immense enterprise.*

J. HOWARD MOORE, *THE UNIVERSAL KINSHIP*

It began while walking along a beach in La Paz, Mexico, in late 1999. My father had died only a few days before, and I expected to spend my time on the beach writing my memories of him. How we came to La Paz is a story in itself. A few days before, my husband, Jim, and I were taking a flight to Los Angeles where my father was to be buried. I told Jim that we needed to rest after the funeral before returning home. I needed some time to reflect on daughter-father interactions before plunging back into my work studying human-dolphin interactions. Jim picked up the airline travel magazine from the seat pocket in front of him and showed me the front cover featuring a glossy map of sunny Mexico. "Maybe Papa wants us to go to Mexico," Jim said lightly but with a solemn smile. "Sure," I replied as my tears stopped for a moment. "Let's just go where your finger lands on the map," I said. The surreal feeling of death—and even our own mortality—called for some irrational behavior.

Fortunately, Jim's finger landed on a location that was easy to fly to. The city of La Paz, meaning "the peace," seemed like the best place possible for some healing. A travel agent Jim found in the phone book made all the arrangements for us, not knowing that we had any interest in dolphins, only that we wanted a quiet beach.

Upon arrival at our hotel in La Paz, we walked out to the beautiful beach in front, margaritas in hand, hoping that the alcohol would serve its intended numbing purpose. Before we even reached the sand, we saw wild dolphins in the distance, their dorsal fins gently surfacing then disappearing as they swam past—healthy, free, and unencumbered. Such a sharp contrast, I thought, to the debilitated and crippled body my father inhabited prior to his passing. In the warm sun we took solace in the beauty of the pelicans flying by and the quiet—except for the water lapping against the sand. The wind that caressed my face brought with it my beloved pelagic ocean winds from which I had been separated for so many months while in hospitals.

Our eyes rested upon a large structure at the end of a pier where some men were working. "Must be an aquaculture site," I said lazily to Jim, feeling naïve in a strange land and content to stay that way. My ever-inquisitive husband, however, had other plans and was compelled, with me reluctantly in tow, to investigate this strange-looking structure. What we found jarred my stupor with a harsh bite of reality. Instead of housing fish, the tiny, shallow pens being built in the stagnant-looking water were to confine dolphins that were to be captured from the wild. With this discovery, the "city of peace" began to transform into a very unappealing place, one that would soon be the center of a heated international controversy. The pain that I felt over the personal loss of my father bled into grief for the planet—the overexploitation of the natural world and the inhumane treatment of sentient beings to satisfy an insatiable desire for money.

The complete and utter irony of my being here—me, a so-called "expert" in the study of swim-with-the-dolphin programs—was not lost on me. Having conducted the first studies of these programs in captivity and in the wild and continuing to specialize in their study, I found it strange to have stumbled upon this discovery by chance, literally "on our doorstep." My amazement and sense of responsibility were the only things that kept me going as Jim and I abandoned our plans for recuperation to investigate this situation. Symbolically in Greek mythology, dolphins have served as "messengers" to help people transition between the worlds of the living and the dead.[1] Only in this regard did it make any sense to find myself thinking about dolphins during this period of mourning.

Jim and I knew we needed to learn more about this facility that was so blatantly inadequate for housing dolphins. The depth of the water in the dolphin pen at low tide was only waist-deep. Shocking to me, but what might this mean to a dolphin? Lack of adequate depth for dolphins can make it impossible for them to move their bodies naturally in and around the water and to exhibit normal behaviors. It can also result in abnormally high water temperatures from which the dolphins cannot escape to deeper, cooler water. It can contribute to stagnation of pollutants in the water in which the dolphins live. And it can affect the transmission of sounds in the water, especially when the water is adjacent to the shoreline. This is a particular concern when there are numerous powerboats traveling so close to the pens. As it turned out, these problems were only part of what we would discover later.

We returned home a week later during the December holiday season. Now, having lost both parents and having no siblings, I felt oddly misplaced during this time of family unity. However, I found myself persistently and lovingly enveloped by my husband and friends who treated me as if I had been a member of their families for years. I thought about the simultaneous fluidity and stability exhibited by some bottlenose dolphins in their interactions with one another. "Fission-fusion" is the term often used to describe this complex type of sociality.[2] In particular, I thought about how dolphins have been known to come to the aid of other dolphins, even those not genetically related.[3] I found myself appreciating how dolphin-like my loved ones were during this otherwise isolating time of life in which I missed having a natal family.

The holiday season also brought with it an announcement that eight bottlenose dolphins had been captured for the La Paz facility. They had been transferred to the facility in the dark of New Year's Eve from Magdalena Bay where they were originally captured. We were shocked that they had been captured so quickly; we had been told that it would not occur until spring. Still grieving the loss of my father, it was easy to imagine what it might be like for dolphins to be separated from their companions and family. If the dolphins felt similar feelings of grief, how could they possibly demonstrate this in a way that these people could understand, let alone care about?

THE INVISIBLE DOLPHIN HEART

I reflected on my study of behavioral indicators of "internal states," as scientists call them, in dolphins. It would be simpler to use the word "emotions," since that is often what we really mean. However, my scientific training has taught me to dance around this word in the discussion of nonhuman animals. But the more we learn about animals, the more compelling evidence of complex emotional lives is discovered and the more awkward this dance becomes. It is not unreasonable to predict that, with growing evidence of sophisticated cognition, tool use, and culture in various species of cetaceans, they would also experience emotions.[4] As Lori Marino and her colleagues have observed, dolphins "share those neurological, cognitive, and social characteristics with great apes and humans that are generally regarded as having been important for the development of self-awareness in primates."[5]

Just a few months before, I had spoken at a symposium on animal emotion at the Smithsonian Institution, facilitated by the pioneering cognitive ethologist Dr. Marc Bekoff. The speakers expressed our relief at having the opportunity to "come out of the closet," so to speak, and to share with our colleagues compelling evidence of emotions in animals. We concluded that there is no longer a legitimate reason why the burden of proof should not be shared with those who deny the existence of complex emotion in nonhuman animals. As Bekoff stated, "denying emotions to animals because they can't be easily studied does not constitute a reasonable argument against their existence." These words would ring through my mind often over the next few months, especially when the state governor of Mexico later declared that he would only intervene to help the La Paz dolphins if it could be "genuinely proven" that they were suffering.[6]

This reminded me of a time when I, along with two American veterinarians, was hired by the Bahamian government to inspect the three dolphin facilities that were in the Bahamas at the time. One facility was so bad that all three of us recommended it be shut down immediately. Before we arrived, one dolphin had already drowned from getting caught in rope within the dolphin pen. And two of the trainers had contacted me secretly to describe how the dolphins were being "overworked," being forced to swim with tourists for sometimes twelve hours a day. In the short time that I had to observe the dolphins, I noticed that one was isolated by himself in a separate pen from the others, swimming in a single tight circle, over and over and over again. This was classic "stereotypic" behavior, usually associated with stress. McIver, as the dolphin was called, was a male who had been isolated from the other dolphins because he was too "sexually aggressive" with the human swimmers. Only a month or so after the inspection, I heard that he had died. Although I cannot prove that he died from the duress of social isolation, I can say that he died a lonely death.

I had seen male dolphins isolated before at captive facilities in the United States for being sexually "inappropriate" with the guests as well. This is why females are preferred over males for swim programs. This is also one of the reasons why the explosion of captive swim programs around the world are such a potential threat to already depleted dolphin populations. Studies have shown that removing females from wildlife populations can result in particularly harmful consequences to the populations.[7] Regardless of gender, since entire groups are often harassed in the process of removing individuals, capturing even one animal from a dolphin group can result in the physiological stress, injury, or death of other nontarget animals.

FROM OCEANS TO PENS

Stunned that the dolphins had been captured so quickly, many people and I wrote letters to Mexican government officials and the owner of the dolphin facility. I stated that these dolphins would be perfect candidates for a successful return to the wild should they reconsider their plan. There was no legitimate research need for the dolphins to remain captive in such a substandard facility. Especially not just so people could thrill themselves by swimming with them.

Although my letter did not receive a response, it did take on a life of its own without me, finding its way to reporters, animal protection organizations, and international Web sites. Shortly thereafter, video footage showing part of the transport of the dolphins was made available. This landmark footage, taken by Juan Antonio Ramirez, documented the most atrocious and inept transport that I—and many others—had ever seen. It is a textbook example of what not to do—or, rather, how to torture a dolphin. The sight of a dolphin's jaws slowly opening and closing while flailing about in what could only be described as agony as it was dropped and dragged repeatedly during transport to the facility was not something that I needed academic training to interpret.

This videotape offered undeniable evidence of cruelty and incompetence. In no time, activists in Mexico such as Dr. Yolanda Passini (who had already written a report on other substandard dolphin facilities in Mexico), Grupo de los Cien (Mexico's most prominent conservation organization), as well as others around the world mounted a letter-writing campaign to the Mexican government and the facility owners.

On February 3, Luna, one of the eight dolphins captured for the facility, died, apparently from an ulcer and related complications, at the young age of eight years (in the wild, dolphins are known to live past fifty).[8] This was not a shock, given that a recent study found that the risk of mortality for dolphins increases sixfold in the month or so immediately following capture.[9] But the fact that this news was not surprising did not make it any less sad. With Luna's death, the public response to this campaign expanded further and became huge, with a torrent of letters pouring into Mexico. The *Washington Post* featured a front-page piece; *Time* magazine, the BBC, NBC Nightly News, the *New York Times*, and various other prominent U.S. and international television media venues highlighted the controversy. Already, this dolphin swim facility had spurred more negative international media attention than any other I had known.

As the protest campaign grew, the environmental minister's office sent a colleague and me an invitation to conduct an inspection of the facility in La Paz as well other dolphin facilities in Mexico that had also received some bad press. We were asked to assist them in creating what would be new regulations for the maintenance of captive dolphins. Simultaneously, I was receiving a barrage of calls from numerous journalists inquiring about the "exploding" number of dolphin swim programs around the world.

In May, the flood of complaints resulted in government authorities announcing a moratorium on the capture of dolphins from Mexican waters. Shortly thereafter, the authorities temporarily closed the La Paz dolphin facility. The *Toronto Star* reported that "Environmental Minister Victor Lichtinger was 'personally very, very worried' that the dolphins won't survive much longer in 'negligent' conditions in a shallow ocean pen. They can toast to death in the summer, and Dr. Lichtinger knows that. He really hopes we can negotiate their release very soon," said ministry biologist Silvia Manzanilla, who advises Lichtinger on marine mammals.[10]

But as the weeks went on, things shifted quietly but perceptibly in Mexico. My invitation from the government was rescinded, with no explanation. Worse, I was notified that government efforts to rehabilitate and return the dolphins to the wild had been abandoned. Even the Mexican Navy, which had been prepared to loan two helicopters to transport the dolphins, was called off the project.

LOVING DOLPHINS TO DEATH

I did not hear much out of Mexico for several weeks. But enough was happening in other areas of the world to continually remind me that La Paz was only the tip of a rapidly growing iceberg. It seemed like almost every day I heard about a facility in yet another country that wanted to capture dolphins for people to swim with or be "healed" by. Or I would receive e-mails from people concerned about deplorable conditions in already existing facilities. For example, I was told that dolphins who were previously used in a traveling "pet the dolphin" show in Guatemala were languishing where they were "abandoned" in a plastic-lined hole dug in the ground after being confiscated by government officials. (I am happy to say that these dolphins have been rehabilitated and released to the wild due to the efforts of a non-profit organization.) Most recently I learned about more dolphins scheduled to be captured for swim programs on several of the Caribbean islands.

While browsing the Internet one day, I saw the news headline "Mexican Dolphins to Stay Penned."[11] The article read,

> Mexican environmental officials were in shock last night over a judge's decision to reverse last week's ruling by the Environmental Enforcement Agency to shut down the facility in the seaside town of La Paz where seven of the mammals are being held in negligent conditions. . . .
> It is not final. But the judge has 22 months to review the case, which could be too long to ensure the survival of the La Paz dolphins. . . .
> And in an extraordinary move, the environmental minister appealed to the international community to keep up the pressure on the owners of the La Paz facility.

The environmental minister was quoted as saying, "Mexican law defends the owners of these [swim-with-the-dolphins] centres, and not the animals. . . . It's bad law."[12] The feeling of disappointment from Mexico and from others around the world was palpable.

UNDERCOVER BIOLOGIST

A few months later, I received a call from ABC's "20/20" in which I was invited to accompany their people to the La Paz facility for a show on this situation. I was hesitant to return to this place that reminded me of such a sad time in my life. But I returned to Mexico with the show's reporter, Arnold Diaz, and the producer and her assistant.

Having advised the "20/20" folks on what to look for at the facility beforehand, we arrived at the facility with a cameraman. I began an informal inspection while everyone else chatted. I quietly noted many violations of U.S. standards[13] for captive dolphins, including inadequate depth and size of pens, fencing in disrepair with sharp edges sticking out where the dolphins could be injured, a lack of sanctuary area for the dolphins to escape from human swimmers if needed, concerns about water quality, and inadequate management of the dolphins. I noted that no real educational information about the dolphins was given to any of us. I also noticed that some of the dolphins had very visible scars on their bodies as a result of the transport.

I chose not to swim with the dolphins, which Diaz and the producer's assistant did (despite my gory descriptions of people's injuries). With my knowledge of their aggressive/defensive behavior in captivity, I know better than to go in with a dolphin who does not know me and whose behavior I am not familiar with. In addition to the roughly eighteen injuries to swimmers documented by the U.S. government within a five-year period, I personally witnessed many more injuries than those reported during this time. I have even seen people get bitten by dolphins at petting/feeding pools. Swimmers have been hospitalized for internal injuries, serious wounds, and broken bones following "attacks" by captive dolphins. A captive orca (in collaboration with his two tankmates) killed his trainer and was implicated in the death of another person.[14] A free-ranging bottlenose dolphin in Brazil killed a man and hospitalized another in self-defense after being abused by these men, who were also found to be drunk.[15]

Dolphins are not vicious, merely wild. Any wild animal, especially one forced into artificial confinement, must be respected for the animal he or she is—not for how he or she can serve us. When this respect is absent from the relationship, both people and animals can easily get hurt. The very nature of these swim programs prohibits the possibility of developing such a respect. Even Dr. Betsy Smith, a recognized developer of dolphin-assisted therapy who no longer works with dolphins for ethical reasons, stated, "People would never throw their child in with a strange dog, but they'll throw them in with a strange dolphin. What you are looking at is vulnerable people and vulnerable animals."

Needless to say, when the trainer in La Paz assured Diaz that the dolphins would not bite him, it was hard for me not to grimace.

MISUNDERSTOOD DOLPHINS

One of the biggest misperceptions about dolphins is that they are always happy and not dangerous or suffering. The other is that they always want to be near us. Their fixed "smile" can be misleading, even though the dolphin wears it through life and through death. I started observing this when studying captive swim programs in the early 1990s. When dolphins exhibited signs of agitation or stress around swimmers, the swimmers typically did not recognize them or laughed them off and chose to misinterpret them as signs of fun. Even obvious signals used by dolphins to indicate warning or frustration to each other (such as open-jaw threats and charges) would frequently go unheeded. Typically, the swimmers would continue doing what they were doing and then be stunned—even offended—if the dolphin forcefully bumped or bit them. It is no wonder that the only two studies conducted on captive swim programs both observed stress-related dolphin behaviors that may be related to negative physiological states.[16]

Free-ranging "friendly" dolphins who interact with people are frequently harassed, hurt, and even killed by people—intentionally and unintentionally. From the Caribbean to South America to Hawaii and Canada, I have found that most recreational and professional boat operators earnestly swear that they do not harass the dolphins or whales they are watching. However, I have seen many of them (especially in countries in which cetaceans are not legally protected) operate their boats in ways that are highly dangerous for the dolphins, such as pursuing them at high speeds, occasionally even separating mother and calf pairs. When a boat chases dolphins—or when swimmers interrupt a group of resting or feeding dolphins—dolphins may exhibit signs of disturbance, such as a repetitive slapping of the flukes at the surface. However, as in the captive swim programs, more often than not, people will choose to believe this to mean that the dolphins are "happy" to see them or are "giving us a show." A consumer-driven ecotour industry (with notable exceptions) and overzealous desire to be near dolphins can certainly contribute to such errors in judgment and sensitivity.

I have also been susceptible to the lure of the dolphins. Thus I began my career, in part, because of a fascination with interspecies communication.

With Dr. Jane Packard of Texas A&M University, I devised a way to apply quantitative behavioral methodologies to analyze the subtle intricacies of human-dolphin interactions. We saw the study of human-dolphin interactions as no more sensational or mystical than the study of how different primate species interact with one another, or how ravens interact with wolves. We are, after all, one of several species of great apes (along with chimpanzees, gorillas, and orangutans) with whom we share the same animal "family," Hominidae.[17]

We theorized that captive dolphins may expect humans to respond to them as other dolphins would (a phenomenon described as "assimilation tendency" by Dr. H. Hediger).[18] However, the startling results were not found in the communication between the two species but rather in the extent of *mis*communication that was occurring, especially on the part of the swimmers. Although dolphins seemed to alter their behaviors to accommodate their interactions with people, the people seemed less adept at reading and responding to the signals of the dolphins. It was not even that the swimmers expected the dolphins to behave like people. It was as if they expected them to behave like "Flipper." I reluctantly switched my career specialization to the study of interspecies *mis*communication.

THE MYTH AND THE REALITY

Regardless of the realities and risks encountered when swimming with dolphins, I am amazed at the ways in which these animals do occasionally live up to their mythical reputation. There have been numerous accounts, many in recent years and from reliable sources, of dolphins assisting human swimmers in distress by holding them at the surface or by pushing them to shore.[19] There are even recent reports of dolphins protecting swimmers from sharks.[20]

The whaling literature of the eighteenth and nineteenth centuries is replete with examples of altruistic behavior of cetaceans directed toward each other. One would consider these people particularly good sources considering that they would not likely be prone to romanticizing the animals they hunted. A multitude of reports exist of dolphins and whales pressing and biting against the lines of ensnared or harpooned animals, propelling the injured animals away from their captors, remaining with injured animals and supporting them at the surface, and, occasionally, attacking the boats inflicting the injuries on their companions.[21]

Some people have tried to argue that dolphins are not aware of what they are doing when they assist others in distress and consider this a type of "fixed" or "innate" behavior. However, two dolphin biologists documented a case in which an adult pilot whale was shot.[22] As it drifted toward the whaling vessel, two other pilot whales rose on either side of the animal, pressed their snouts on top of its head, and took the animal *under* the surface and away from the vessel, not to be seen again. The animal was not supported in the expected method of being held at the surface, but in a clearly contradictory manner. In another case, the pioneering cetologist Ken Norris personally observed a captive roughtooth dolphin pulling a hypodermic syringe from another dolphin in the tank during medical treatment. He reported that the assisting dolphin then became quite aggressive toward the veterinarian who inserted the syringe in his tankmate.[23]

As I write this, however, I caution anyone against depending on dolphins to save their lives if in danger at sea. I have personally been abandoned by dolphins I was observing underwater in the Bahamas, only to find myself swimming with a shark instead! And I have received infrequent but reliable reports of pilot whales, orcas, and other cetaceans intentionally striking small boats until, on rare occasion, the vessels have been damaged.

When these different behaviors are considered, it appears that one can have a considerable range of experiences while interacting with dolphins, from becoming injured to being rescued. Although this does not serve to perpetuate a mythical stereotype, it does promote a more complex and intriguing portrait of these animals.

PRISONERS OF LOVE

The title given to the "20/20" show when it aired was "Prisoners of Love." A silly title, I thought initially, but it also seemed appropriate. The announcer's voice said, "We're having fun, but what about the dolphins? We bet your vacation video doesn't show you this—dolphins being dragged, dropped, traumatized, and, then, penned up for your pleasure." Pan to Barbara Walters saying, "Sometimes dolphins pay a high price for being so popular."

My interview segment was brief, but I was glad that they included at least one biologist's perspective. It is important that the public know it is not only animal rights activists who oppose these programs, as the media so often and so erroneously portrays. I know, from the workshops that I have led at academic conferences on this subject, that many of the world's top dolphin biologists and trainers have serious concerns about interactive programs with dolphins. However, many biologists do not speak publicly of this out of fear of losing their jobs or grant money. For example, on my first trip to La Paz I met with two biologists in Mexico who worked for the government. They both told me that the dolphin facility was not suitable for dolphins, and one expressed concern about the wild population from which these animals were captured. However, they both asked that I not disclose their names on this matter, and they remained silent during the public deliberations.

The only important topic that the show failed to cover, in my opinion, was that of the educational value (or lack thereof) of dolphins in interactive programs. Some people think we need to see dolphins in tanks so that we will be motivated to protect them in the wild. However, I have heard it said that the "save the whales" movement was possibly the largest animal conservation movement in the world, and it was accomplished without people seeing whales in captivity, let alone swimming with them. I am not aware of any peer-reviewed research that demonstrates the educational value of dolphin swims, or captive dolphins in general, to the public. If anything, I have come to believe that a false and potentially harmful message is imparted to the public by displaying wild animals, especially marine mammals, in captivity. It teaches people by example that they are not inextricably linked to their natural environment, where they have evolved and to which they belong. As Trevor Spradlin of the National Marine Fisheries Service told me, "There is growing concern that feeding pools, swim programs, and other types of interactive experiences with marine mammals in display facilities may perpetuate the problem of the public feeding and harassment of marine mammals in the wild, especially if they do not educate their guests to respect wildlife." As Dr. Naomi Rose stated, "the media's glamorization of dolphins has led people to believe that their interactions with them are benign, when in fact they can be extremely disruptive."

After years of watching these programs, to me they look like little more than glorified petting zoos, using exotic dolphins instead of barnyard domestics. I doubt that most people will be any more inspired to work for marine animal protection after participating in these programs than people will become vegetarians after visiting a petting zoo. Not only does the public not learn much about the lives of these animals from watching them in captivity, but they also go home thinking that the tricks they saw the dolphins perform are really how dolphins behave in the wild. They are more likely to leave thinking of dolphins as playthings or toys. However, the perception of dolphins as "playthings" can even occur when people interact with cetaceans in the wild. This was made poignantly clear to me when I overheard a man with his son out on their boat, looking for a beluga whale with whom they had previously interacted by shouting out, "Where's our toy?"

PROGRESS

At the time this book goes to press, it appears that the dolphins in La Paz will remain in their pens and their owners will continue to capitalize on their misery unless something remarkable happens. Yet Dr. Passini and others in Mexico and elsewhere continue to work for their freedom, and this chapter may never be truly completed. However, the publicity over the loss of one dolphin's life and the suffering of the others there appears to have resulted in a radical shift of legislation governing all dolphins in Mexico. The Mexican government has since approved a set of standards governing the care of captive dolphins that, if enforced, may at least moderately improve the welfare of the La Paz dolphins and others in Mexico. Even more far reaching is the passage of laws that prohibit the hunting or capture of any species of marine mammal for commercial or subsistence purposes, including export to other countries. In fact, Mexico has just become the world's largest whale sanctuary, offering protection to over thirty-nine cetecean species in nearly 1.15 million square miles of Pacific, Atlantic, and Caribbean waters. The significance of this legislation is enormous in that marine mammals in Mexico were previously afforded only the most minimal protection. All of these laws require enforcement, but the fact that they have been enacted, especially all within a year's time, is nothing less than monumental. It also demonstrates how efforts to protect individual animals can translate into protecting entire populations and ecosystems.

In December of 2002, a colleague (with the Humane Society of the United States) and I traveled to the Dominican Republic at the invitation of that nation's Academy of Sciences and the Fundacion Dominicana de Estudios Marinos. Scientists and others in that country were concerned about the impacts of captures on the local dolphin population. As in the case of La Paz, a wealthy foreigner had recently captured eight dolphins (in addition to others previously captured for a substandard facility) for use in swim programs. However, the Dominican Republic had recently designated the waters from which the dolphins were captured as a marine sanctuary, and a law had been enacted that prohibited their removal from the wild. These laws were milestones for people who had only recently been able to apply their newly found democracy to environmental issues, following years of political oppression. Now they wanted to see these laws enforced.

The dates chosen for our visit happened to coincide with the anniversary of my father's death. On the last evening of this trip, I found myself visiting with an amazing group of people—including the mayor and town council members—in the small seaside village of Bayahibe, which borders the marine sanctuary. Here, locals had been familiar with the dolphins for many years and had made great efforts to see the laws enforced to protect them. "Why do the local people here care so much about the dolphins?" I asked. My question was met with beaming eyes and smiles. "The dolphins escorted the painting of *La Divina Pastora de las Almas* [the "Virgin Mary" of this village] in the annual procession of boats during her festival in May. We had never seen anything like it before! Artists will even paint a mural in town to commemorate this event" I was told. The people's enthusiasm conveyed nothing less than a belief that they had witnessed a miracle. I felt deeply privileged to hear this story of union between spirit and nature, especially on such a sacred day for me. I remembered how the birthday of *La Virgen de Guadalupe*, the Holy Virgin of Mexico, was being intensively celebrated throughout La Paz during my original visit there.

The efforts of the citizens of the Dominican Republic to protect these animals—including an impending lawsuit as this book goes to press—are among the first formal efforts to enforce environmental laws in this country. So the fight to free and protect these dolphins is intertwined with a fight for human rights and personal and political empowerment. And again, the efforts to protect eight dolphins have transformed into something much larger than the dolphins themselves.

FREEDOM AND FAMILY

Educators say that people protect what they love. But if we love dolphins so much, how do we justify taking from them the very things we value most in our own lives: freedom and family?[24] If the bond between humans and dolphins is so sacred, then how do we justify exploiting it in such harmful ways? I cannot pretend to know what dolphins feel, but I cannot help but think what a cheap substitute human swimmers and frozen fish would be for the freedom of the vast, unpredictable ocean and the company of myriad dolphin companions and family.

The aftermath of death (human or otherwise) can bring with it a greater appreciation for life. The passing of my father and Luna, a dolphin whom I had never even seen, each affected me very differently. But each loss inspired the honoring of life and, in this way, it may be said that neither died in vain. My grief over my father transmuted into a renewed commitment to contribute to the lives of others. Luna's death ignited an international firestorm of compassion resulting in sweeping legislation to protect dolphins. And without my knowing it, trying to help the La Paz dolphins contributed to my own healing. Although the myths tell of dolphins assisting in the transition from life to death, in my case they were a part of my reintegration from the world of the dying to that of the living. They served as a powerful catalyst in my transition from a state of separation to one of profound awareness of my interconnectedness to all life.

I am not sure why humans have such a hard time leaving wild animals to the wild. It is as if we have forgotten what it is to be wild ourselves. Perhaps we need to reconnect with what little has not been domesticated within ourselves before we destroy all that is wild around us. To "love" dolphins in a way that does not injure them requires interacting with them on *their* terms rather than ours. This may involve passing up opportunities to be with the very animals that inspire us the most. But by protecting what is wild in other animals, we may also salvage the wild in ourselves. What seems to be a personal sacrifice in the short term may, in the long term, translate into an even greater connection with our own animal nature—and, ultimately, our extended animal family.

NOTES

1. See Pat Weyer, "Signs of Becoming" (in this volume).

2. R. C. Connor, R. S. Wells, J. Mann, and A. Read, "The Bottlenose Dolphin," in *Cetacean Societies: Field Studies of Dolphins and Whales*, ed. J. Mann, R. C. Connor, P. L. Tyack, and H. Whitehead (Chicago: University of Chicago Press, 2000).

3. R. C. Connor and K. S. Norris, "Are Dolphins Reciprocal Altruists?" *American Naturalist* 119, no. 3 (March 1982): 358–74.

4. L. Rendell and H. Whitehead, "Culture in Whales and Dolphins," *Behavioral and Brain Sciences* 24 (2001): 309–82.

5. L. Marino, D. Reiss, and G. G. Gallup, Jr., "Mirror Self-Recognition in Bottlenose Dolphins: Implications for Comparative Investigations of Highly Dissimilar Species," in *Self-Awareness in Animals and Humans: Developmental Perspectives*, ed. S. T. Parker, R. W. Mitchell, and M. L. Boccia (New York: Cambridge University Press, 1994).

6. *Toronto Star*, June 13, 2001.

7. M. Oldfield, "Threatened Mammals Affected by the Exploitation of the Female-Offspring Bond," *Conservation Biology* 2, no. 3 (1988): 260–74.

8. Connor et al., "The Bottlenose Dolphin."

9. R. Small and D. P. DeMaster, "Acclimation to Captivity: A Quantitative Estimate Based on Survival of Bottlenose Dolphins and California Sea Lions," *Marine Mammal Science* 11 (1995): 510–19.

10. *Toronto Star*, May 30, 2001.

11. Ibid., June 7, 2001.

12. Ibid., June 13, 2001.

13. Animal and Plant Health Inspection Service, U.S. Department of Agriculture, "Subchapter A—Animal Welfare; Subpart E—Specifications for the Human Handling, Care, Treatment, and Transportation of Marine Mammals," Sec. 3.100–3.118 in Code of Federal Regulations, Title 9 (Washington, D.C.: Office of the Federal Registrar, 2001).

14. *Edmonton Journal*, May 5, 1991, and Reuters News Service, Sept. 20, 1999.

15. M. C. O. Santos, "Lone Sociable Bottlenose Dolphin in Brazil: Human Fatality and Management," *Marine Mammal Science* 13, no. 2 (1997): 355–57.

16. T. G. Frohoff, "Behavior of Captive Bottlenose Dolphins and Humans During Controlled In-Water Interactions" (Master's thesis, Texas A&M University, 1993); A. Samuels and T. R. Spradlin, "Quantitative Behavioral Study of Bottlenose Dolphins in Swim-with-Dolphin Programs in the United States," *Marine Mammal Science* 11, no. 4 (1995): 520–44.

17. D. Wilson and D. Reeder, eds., *Mammal Species of the World: A Taxonomic and Geographic Reference* (Washington, D.C.: Smithsonian Institution Press, 1993).

18. H. Hediger, *Wild Animals in Captivity* (New York: Dover Publications, 1964).

19. "Dolphins Said to Have Rescued Fishermen," Reuters, October 25, 1993.

20. For example: "Dolphins to the Rescue: Fellow Mammals Repel Shark, Save Surfer," *Surfer Magazine*, May 1993; "Dolphins Save Man in Shark Attack," *Washington Times*, July 26, 1996; "Columbia Stowaways Say Dolphins Saved Their Lives," Reuters, February 18, 1999.

21. Connor and Norris, "Are Dolphins Reciprocal Altruists?"

22. Ibid.

23. Ibid.

24. In a 1998 presentation in Seattle, Ben White discussed a perceived conflict between the American values of freedom and family with the capture and confinement of captive dolphins in the United States.

The Dolphin's Gaze

One brief experience thirty years ago charted the entire course of my life. It was the morning I first swam with the wild dolphins in a remote bay in Hawaii. I am still trying to figure out exactly what happened. All I know is I was a different person when I crawled from the sea than the one who plunged in an hour before. Colors were brighter. Gratitude for being alive moved me to a strange combination of laughter and tears. My old worldview had collapsed on itself. A new one had sprung from the sea, from my swim, and from the dolphin's gaze.

I am not a "new-ager." As a lifelong professional arborist and tree climber, as well as a longtime single daddy, I love solid and sure things, like the steel snap of a climbing clip telling me I am safely tied in. I believe in the tangible. But my source of sustenance has always been a personal relationship with Nature. As a kid, I found solace exploring creeks and forests.

When others of my generation scattered across the globe searching for spiritual teachers in the Himalayas or in the zendos in Japan, I wound up—at twenty—living by myself on a jungle hillside on the big island of Hawaii. Three or four times a week I would follow the winding asphalt road down to the bay, kick off into the warm clear water, and enter the dazzling world of reef life. In the evening I climbed, dripping wet, back home to my little plastic and bamboo tipi tucked beneath two coconut palms that clattered in the breeze. Back then, this bay was a quiet place and very different from how it would be years later when hordes of people would flock there in commercialized tours to swim with the dolphins.

I meditated and studied Zen Buddhism and Lao Tzu. I was intrigued by the concept of enlightenment and collected odd stories that described the onset of this flip-flop of consciousness. One told of a solitary monk who attained enlightenment after decades of concentration. While sweeping his secluded cabin, a pebble thrown by his broom thwocked against the wall with a peculiar sound. Bingo! He was changed forever. Other stories told of realized beings that could trigger enlightenment with just a glance.

I assume that I have not passed that elusive golden doorway into enlightenment. I still get moody and depressed and bark at my kids. Only twice in my life have I experienced a gaze that really reached to the bottom of my soul and turned me inside out. Once was from a yogi I met in southern Oregon in a cave behind a waterfall. The other was from one of the dolphins in the bay.

For a couple of weeks I had been watching dolphins glint in the sun as they played in the bay. Viewed from my hillside high above, one after another would crash through the ocean surface and spin high into the air with the most amazing pirouettes before splashing down. The joy and abandon in their movement beckoned to me. I wanted to play. Inviting a few laid-back friends to join me on the mile or so swim out to the dolphins, I found no takers. When I asked why the dolphins kept flying out of the water I was solemnly told that it was their way of shedding parasites. Sort of like saying Baryshnikov soared due to ants in his pants.

A dozen excuses argued I should not swim out to the dolphins. Finally the desire for adventure overwhelmed the fear of risk. Sitting at the surf's edge, just barely able to see distant dorsal fins slice the surface, I strapped on mask, snorkel, and fins, lowered my head, and started kicking. The water world was familiar, the moving reef life a comfort, until it began dropping away. Hawaii is just the tip of the largest mountain in the world; its base is thousands of feet underwater. I was tiny, almost naked, alone, flying over this abyss until I could no longer see the bottom, just light shafts from the sun not quite converging far below in the indigo depths. A giant barracuda angled toward me, a flashing silver bullet. But he passed by, heading toward the shore, without any sign of notice from his flat round eye.

On and on I kicked. Now and then I raised my head to check my progress. Thoughts raced: Yep, they're still there, dorsal fins mimicking the short choppy surface waves. But jeez, they're big, much bigger than I thought. Maybe this little swim isn't such a good idea after all.

I started to get scared, but it was too late to retreat. They had spotted me. Two dolphins suddenly shot straight toward me. I heard an odd trilling series of clicks and felt a ratcheting vibration shudder through my chest and belly. I had been echolocated.

One dolphin kept coming at me. I wondered if dolphins really did kill sharks by ramming into their bellies like in the show *Flipper*. There was no place to hide. I was totally out of my element, defenseless. Then the dolphin slid by on my left, maybe eight feet away, and *regarded* me from stem to stern with one long piercing gaze.

In trying to understand this moment and why it triggered such an epiphany, I keep thinking about that gaze. We are told that eyes only receive, not send. I do not believe that. My favorite game as a bored kid in school was to stare at the back of another student. Invariably they would turn around. Something comes *out* of eyes.

What happened when that dolphin looked at me? I was humbled to the dirt. All of my insecurities and fears were shaken out like so much dust; inspected, laughed at, and discarded. The gaze stripped me of all swagger and presumption.

But it was more. I said the dolphin "regarded" me. I have known many dogs; close friends I loved dearly. But I have never seen such complexity, humor, and *recognition* in the eyes of any creature other than humans, and rarely enough in those. Inside that sleek gray dolphin body was a person. No doubt about it: a self-aware, evaluating, conscious, thinking, playful, and accepting person.

If dolphins had "persons" inside of them, then almost everything that I had been taught about human specialness and our perch upon the crown of creation was a crock. A whole string of logical assumptions tumbled like dominoes. If dolphins are persons, then all other creatures must also be persons, even if their eyes do not shake my innards quite the same way. And if all other creatures were persons, self-aware entities like me, then the world was infinitely richer. The axiom that all other creatures except for humans were just inert props for our starring role on earth was exposed as a deadly lie, keeping us isolated from our greater family. The world was suddenly huge and welcoming, with every facet calling out to me. I had found a place I belonged. It replaced the sterile stage upon which I was briefly performing.

The guard dolphin let me pass. Soon I was surrounded by a pod of about fifty dolphins. (Years later I learned that this family, now much besieged by seekers just like me, was one of only two resident pods of spinner dolphins known at the time.) All around me swam old scarred bull dolphins, little babies snug alongside their mothers, and mating dolphins belly to belly. Frisky young dolphins raced under me before exploding through the sea's silver ceiling, only to whack down seconds later wreathed in rainbow bubbles.

The dolphins moved with so little apparent effort, as if they were watermelon seeds squirted by invisible pinched fingers. In contrast, I felt sillier than a fish out of water. I was a man off land; goofy, elbows and knees hanging down while superior beings showed me what individual movement could be.

Too soon, the dolphins vanished. No more symphony of squeaks, clicks, and whistles. They simply disappeared. I could not even see them when I lifted my head to peer above the chop. I yearned to follow, to join their world, to learn a little more. But I had to go back to land, to people.

Kicking the mile back to shore, the world was brand new and in Technicolor. A sense of obligation followed directly on the heels of euphoria. I had been given a glimpse into the world of the Other. It had cracked me open. These creatures needed nothing from me. They were complete. But all over the world, human beings were hurting dolphins and whales. They were not able to speak in their own defense. Why couldn't I, as a human being, try to speak their voice within the world of human laws and practices? I made a solemn vow to use my life to protect these creatures from my own species.

That vow has, over almost three decades now, led me into over fifty countries and many foolhardy missions. It took me to Japan, Mexico, the Bahamas, and Florida to cut nets to free captive dolphins and whales. It has led me to many interminable meetings of the International Whaling Commission, the Convention on International Trade of Endangered Species, and the Inter-American Tropical Tuna Commission, where the stroke of a pen can doom or save thousands of individuals.

Of course, those who kill or capture whales and dolphins do not believe that their victims are "persons." They certainly do not believe that I was recruited to the dolphin's service that day out in a remote Hawaiian bay.

Their disbelief does not matter, because the dolphins' gift to me on that morning has been so pervasive and enduring. I was brought to understand that this earth is peopled with millions of nonhuman conscious entities weaving a musical, sexual, emotional matrix of diverse form and mind. This discovery has transformed my world into a miracle renewed each day. My final gleaning from the dolphin's gaze is this: not only does our greater family of life on earth wish us well, but they have also been waiting forever and a day for us to join in the dance as full participants instead of as lonely paper-tiger bosses.

I am still a wallflower at this dance, just starting to get acquainted and learn the steps. But the simple fact that I have been invited to the party fills me with delight.

JOHN C. LILLY

Toward a Cetacean Nation

To some people, Dr. John Lilly was a creative genius and visionary. To others, he was a scientist gone widlly astray. One thing that is probably agreed upon is that he was an intellectual pioneer—one who has set precedent in the exploration of the dolphin (as well as the human) mind. Although he frequently and unabashedly violated the rules of acceptable scientific practice and behavior, Lilly is arguably considered to be the "father" of the modern-day study of human-dolphin communication (and, in a broader context, interspecies communication). We will never know how many scientists and naturalists were inspired by Lilly's work. I am sure that all of the contributors in this anthology have been influenced in some way by his work, regardless of their opinions of his contributions.

Through his research, Lilly enhanced the perception that dolphins were sentient beings, but his methodologies soon radically diverged from those of his peers. He described dolphins as "extraterrestrial intelligence on earth" and espoused prophecies of communication between dolphins and humans without the data to support his claims. While this discouraged the scientific community, it inspired the general public with many books written on the subject. Lilly even served as the model for the movies Day of the Dolphin *and* Altered States.

Over the years, Lilly had studied dolphin intelligence and human-dolphin communication in a variety of highly creative if controversial experiments. But with the ailing health of Toni Lilly, John's wife and research partner, their last project was abandoned before the desired results were achieved. However, prior to Toni's death, she was committed to providing for the rehabilitation and release to the wild of their two captive dolphins, Joe and Rosie, who were used in this research. This rehabilitation, featured in a National Geographic documentary, represented the first documented rehabilitation and release of captive dolphins to the wild.

We are honored to feature what appears to have been the last piece that Dr. Lilly wrote about dolphins prior to his passing in 2001. We are thankful that two of his sons (Philip Hansen Bailey and Charles Lilly) assisted in completing it. We are not surprised that, in his piece, Lilly advocates the need for nothing less bold than a "Cetacean Nation." Of the various things that Lilly has been criticized for, he may have been most reprimanded for the very thing that made him such a significant figure—for boldly taking mythology into the laboratory and then into our every-day lives. Regardless of his methods, Lilly laid the foundation for making the pos-sibility of interspecies communication accessible to people and providing hope that what Loren Eiseley calls our species' "long loneliness" may someday be solaced. For this, John Lilly will always be remembered. — TF

The rights of man have slowly developed over the past few centuries. The rights that each of us enjoy today in the United States and throughout the world have been carefully developed in our laws. Our history teaches us that our rights evolved through certain stages, from the unconscious acceptance of a lack of rights, to the conscious awareness of the need for an adequate expression on an intolerable situation or state, to demand for the relief articulated in law, to the law and its adequate ad-ministration. As each group of humans, through its own experience, learned to feel its lack of sharing in the benefits of laws and their administration, each developed adequate spokesmen and spokeswomen for its cause.

As individual humans and groups, it is we who must become the legal guardians or protagonists for the environment and animals. Henceforth, I will present my argument and proposal for the foundation of a global Cetacean Nation. Recognizing that dolphins and whales have the largest brains on the planet and are quite capable of speech, Cetacean Nation re-alizes their voices must be heard and is dedicated to establishing under-standing between humans and cetaceans.

The intelligence of dolphins has been the subject of speculation by writ-ers since ancient Greece. The discovery of the large brains of the Cetacea, in the eighteenth century, led to inevitable comparisons of these brains to those of the humans and of the lower primates. The winds of scholarly opinions concerning the whales have anciently blown strongly for high intelligence but, during later centuries, shifted strongly against high intelligence. At the time of Aristotle (384–322 B.C.), the dolphin, for example, was held in high esteem, and many stories of the apparently great abilities of these animals were current. By the time of Plinius Secundus (A.D. 23–79), the beginning of a note of skepticism was introduced.

In the Middle Ages, the strong influence of religious philosophy on thinking placed humans in a completely separate compartment from all other living creatures, and the accurate anatomy of the whales was neglected. For commercial reasons, man exploits the bodies of these beings; historically, our most common relationship to the whale has been usury. It was not until the anatomical work of Andreas Vesalius (1514–64) and others that the biological similarities and differences of human and other mammals were pointed out. It was at this time that the investigation of the large and complex human brain began.

All through these periods, intelligence and the biological brain factors seemed to be completely separated in the minds of the scholars. For ancient Greeks and Romans, there was little, if any, link made between brain and mind. Scholars attributed special human achievements to factors other than excellence of brain structure and its use.

After the discovery of the complicated and complex human brain and the clinical correlation between brain injury and effects on human performance, the brain and mental factors began to be related to one another. As descriptions of the human brain became more and more exact and clinical correlations increased sufficiently in numbers, new investigations on the relationships between brain size and intelligence in *Homo sapiens* were started.

By the year 1843, the size of the brain of whales was being related to the total size of the body. The very large brains of the large whales were reduced in importance by considering their weight in a ratio to the weight of the total body. Drawings from 1845 show the improved awareness of the complexities of these large brains in regard to cerebral cortex, the cerebellum, and the cranial nerves. Porpoise and dolphin brains were described as nearly in the same proportion as the human brain. However, intelligence was still not related to the largest brains on the planet. Communication and the behavior of cetaceans were still described as instinctual.

To ask about the intelligence of another species, we first ask: how large and well developed is its brain? Somewhat blindly we link brain size, a biological fact, to intelligence, a behavioral and psychological concept. We know, in the case of our own species, that if the brain fails to develop, intelligence also fails to develop.

How do we judge in our own species that intelligence develops or fails to develop? We work with the child and carefully observe its performances of common tasks and carefully measure its acquisition of speech quantitatively. We measure (among other factors) size of word vocabulary, adequacy of pronunciation, lengths of phrases and sentences, appropriateness of use, levels of abstraction achieved, and the quality of the logical processes used. We also measure speed of grasping new games, with novel sets of rules and strategy, games physical and/or games verbal and vocal.

Consider cetacean brains and our brains.

How are we, in the structure of our brains, different from our nearest relatives, the apes (chimpanzee, gorilla, and orangutans)? The result of careful neuroanatomical and extensive neurophysiological studies indicates that our brains' only decisive difference from the apes is in the size of our cerebral cortical "silent" areas on the frontal, parietal, and temporal lobes.

Silent areas have no direct-input connections and thus are devoted to central processing (thinking, imagination, long-term goals, ethics, etc.). Persons in whom these areas are lost become here-and-now beings, with a good deal of what is typically considered an important aspect of their humanness gone. They lose their moral and ethical judgments and their motivations for future planning and action. These cortical areas are the ones we use for understanding justice, compassion, and the need for social interdependence. If we could control the growth of these cortical areas in our species, we could possibly evolve further beyond our present horizons.

Apparently, we are limited by this aspect of our brains' structure. This may make us pause when we consider the brains of Cetacea. Excellent controlled neurological studies of cetacean brains show that the silent areas are larger in the cetaceans, whose brains are bigger than ours. In the bottlenose dolphin, with a brain 40 percent larger than ours, in *Orcinus orca*, with a brain three to four times the mass of ours, and in the sperm whale, whose brain is the largest on this planet (six times the mass of a human brain), all of the additional mass is in the silent areas. Microscopic analysis shows that their cellular densities and connections are quite as large and complex as ours.

The belief that humans are the preeminent thinkers, doers, and feelers on this planet is denied by these investigations. The brains of cetaceans may be superior to ours, in unique and effective areas. Let us try to find out from the Cetacea their criteria for intelligent use of large brains, rather than attempting to impose our criteria upon them.

I am espousing a plea for an open-minded attitude with respect to cetaceans. It would be presumptuous to assume that we, at the present time, can know how to measure their intelligence or their intellectual capacity. Comparing a handed mammal with a flippered mammal, each of which lives in an entirely separate and distinctive environment, is a very difficult intellectual task even for *Homo sapiens*.

It is necessary, in order to evaluate the intelligence of the dolphins, much less the whales, to know something of their abilities in the areas of phonation and other kinds of bodily gestures and manipulations and, hence, in their abilities to communicate with one another. As I implied in *Man and Dolphin*, it is not possible to measure accurately the intelligence of any other creature than that of a human being, mainly because we do not exchange ideas through any known communication mode with such creatures.

Because cetaceans live immersed in an acoustic world quite strange to us, we have great difficulty in appreciating the full life of these beings with respect to one another and their environment. From birth, they are constantly bombarded with signals from the others and by echoes from the environment. Their ultrasonic (to us) emissions are not merely sonar; they use their ultrasounds and their high-pitched sounds interpersonally with fervor in practically everything they do. A physical analysis of such vocal transactions shows them to be as complex in many of the same ways as the vocal transactions between human beings.

One question often posed is, "If there is extraterrestrial intelligent communicating life, how will we recognize it and how will we communicate with it?"

Currently, we are faced with other species possibly as intelligent as we are. We do not yet recognize their intelligence. We do not even attempt communication with them. We do kill them, eat them, and use their bodies as industrial products. We have no respect whatsoever for their huge brains. In other words, as a species, human beings faced with other species demean them, kill them, and eat them. I hope all this carnage can be stopped, and something more ideal can take its place. The slaughter of these magnificent beings must end. Instead of slaughter, let us devise new projects with whales, conserving them.

Large private resources and government resources should be devoted to encouraging bright and intelligent human beings to devote their lives to this problem of achieving communication with these magnificent brains and minds. I visualize a project as vast as our space program, devoting our best minds, our best engineering brains, our vast networks of computer people, materials, and time on this essentially peaceful mission of interspecies communications, right here on this planet. It is a bit discouraging to see this magnificent opportunity being thrown aside by the human race. I hope that enough people can be fired with the enthusiasm, which I feel for it, to launch the effort with momentum similar to that we have devoted to oceanography, nuclear energy, and warfare. The rewards will be great: new experiences, new ways of thinking, and new philosophies.

We do not apply the Golden Rule to the dolphins or the whales or any species other than our own. Even dealing with our own species, we tend to annihilate vast numbers of them. As a species, we are a poor example for highly intelligent life forms, from other planets or from our own planet, to communicate with. Let us improve ourselves. The human species is so arrogant that it does not recognize its own superiors. The only way that humans as a whole will respect any other species, apparently, is the ability to beat them in warfare. Perhaps the dolphins are applying the Golden Rule to us. In general, whales and dolphins have withheld using killing force with humans.

Someday, we may acquire enough information to discuss the dolphins' real ethics. Our present military potential is great enough to kill not only all human beings but all life forms on the land and in the sea. It may sound a bit silly to go beyond this potential for cataclysm, but modern military minds are carrying the power to create disaster even further. *Homo sapiens* are fast destroying this planet as a habitable abode for all of life as we know it.

It may be that this is the natural course for our form of life. It may be that it is destined for this type of self-destruction. It may be that the evil, in enough of us, is destined to destroy the whole of us. I sincerely hope and pray that this is not so. I hope that my children and their children have the same opportunities to pursue beauty, goodness, truth, and self-fulfillment that I have had.

It may be that there will be no time left for humans to prepare to really communicate with the dolphins and the whales. It may be that we are all doomed to extinction before such an ideal can be realized. Without dedicated people who persist in scientific research that is oriented toward an idealistic set of aims and goals, we may as well give up the ghost as a progressive species and sink into a new dark age.

Sometimes, I feel that if man could become more involved in some problems of an alien species, he may become less involved with his own egocentric pursuits and deadly competition within his species, and become somehow a better being. The whole philosophy that says one species must rule the other species has been cast out of the thinking of myself and my colleagues. We are often asked, "If the dolphins are so intelligent, why aren't they ruling the world?" My very considered answer is—they may be too wise to try to rule the world.

The question can be easily turned around. Why does man or individual men want to rule the world? I feel that it is a very insecure position to want to rule all of the other species and the vast resources of our planet. This means a deep insecurity with the "universes" inside of oneself. One's fears and one's angers are being projected on others outside of oneself; to rule the world is, finally, to rule one's inner realities.

What behavioral evidence do we have of ethics among Cetacea? The main evidence is derived from our encounters with them, and some from our observations of their interactions with one another. In my books (*Man and Dolphin* and *The Mind of the Dolphin: A Nonhuman Intelligence*) are many examples of intelligent and compassionate actions toward us and between individual dolphins.

It is my deep feeling that, unless we work with respect, with discipline, and with gentleness with the dolphins, they will once more turn away from us. If we think of dolphins as "lower animals," then we will not even attempt to meet them as "equals" worthy of our efforts. If we think of them as "bright, intelligent animals serving man" in entertainment, in circuses, and on television, we are favoring a segregationist point of view. Currently, we place dolphins "over there," with chimpanzees, performing dogs, and fictional horses that talk to television audiences. If we "teach" them to aid our underwater work, in the sea, as glorified "seeing-eye dogs" of the Navy, of oceanographers, and of divers, we are far from my goals for them. This sort of a relationship ("fetch and carry") is not between equals.

It may be that the word "equal" is not quite the term to use for dolphin-human comparisons. A dolphin can be what I call a "cognitive equal" with a human being and still be an alien and strange mind, as seen by the human.

Slavery (man to man) has had a long history in our species. Persons can still "buy" dolphins from other persons even though they cannot "buy" another human being. So long as the legal view of dolphins is that of "animals," they will still be an article of commerce to be exploited for man's purposes.

We must somehow translate the Golden Rule, and the dolphin will have to do the same, for use between such strange, likeable, cooperative beings. As one of the "others" in the Golden Rule, neither his appearance nor his alikeness of thought should deter us or him. Both sides will have to search for the bases of equality. "No matter the differences between species, no matter differences of anatomy, no matter differences between media in which they live, creatures with a brain above a certain size will be considered 'equal' with human."

Such considerations inevitably lead us to the conclusion that the Cetacea are a special case. We must give them rights as individuals under our laws. Here are some guidelines for a developing relationship based on respect:

1. Cetaceans are no longer to be considered as property, nor as an industrial resource, nor as stocks of animals.

2. Cetaceans are to have complete freedom of the waters of the earth.

3. Our best intelligence shall be used in determining the needs, concerns, and well-being of Cetacea.

4. Communication between humans and cetaceans shall be encouraged rather than be constrained. Research is to be initiated, encouraged, and supported to establish means of communication with the cetaceans.

5. As communication is established, communications shall be used openly to promote the well-being of Cetacea.

6. New human understandings are to be anticipated and recognized in global affairs, such as in the decisions of the United Nations. Laws shall be proposed to the United Nations based upon equal representation of humans and cetaceans. Thereby recognizing a new Cetacean Nation.

It is time to recognize that the human species has maintained a human-centered, isolated existence on the planet Earth because of its failure to communicate with those of comparable brain size existing in the sea. The cetaceans have a reality separate from human reality. Their reality, defined in their own terms, their social competence, their survival for the past fifteen million years are to be respected and be researched, and the consequences are to be legislated into human law.

There are those who will think of the above proposals in terms of a science fiction script, thus rendering themselves safe from taking these ideas seriously. The answer to such a viewpoint is that this merely shows ignorance of the facts of brains and of ecosystems on this planet. One can safely disavow becoming involved in this very large program by such beliefs. Humans have a long history of espousing blinding beliefs that bind them to ideas that have led to the demise of whole cultures. Opening one's eyes to the possibilities of nonhuman communicators, with immensely complex and ancient histories and ethics on this planet, requires one to shed the blinding beliefs inherited from the past.

Therefore, let us expand our scientific horizons and recommend and support research leading to the testing of our communication capacities with the largest brains on this planet. Let us at least explore the possibility that they are capable in ways that we, in our present ignorance, cannot yet know about. The least we can do is to stop killing them, and the most that we can do is to dedicate the best of our science to testing them and seeing if it is possible to communicate with them. One possibility is that, even with our best methods, we may not be bright enough, or our science may not be adequate, to break the communication barrier. But let us try to see if we can do it. We do need some outside-the-human-species input, we do need the perspective of someone else on our activities, we do need the exercise of negotiating with others, besides humans. We do need to know the ancient wisdom of the dolphins.

PART II

The Many Worlds of Dolphins: International Waters

D olphins live in a fluid world beyond human maps or boundaries. Their own cetacean cultures seem to be as diverse and fascinating as our own. Here we look at our varied cultural traditions and kinship with dolphins around the world.

Our journey begins with "Talking to Beluga," in which musician and naturalist Jim Nollman takes us on an expedition of music and art to the Karelian petroglyph sites of Scandinavia and Russia, where Finno-Ugric tribes carved drawings of cetaceans thousands of years ago. Next, in "At Home with the Dolphins," scientist Giovanni Bearzi shares his studies and experiences with solitary, sociable dolphins and groups of dolphins encountering humans in the seas around Italy.

The Amazon River is the setting for an essay by author/naturalist Sy Montgomery, who offers us a dream-like glimpse of the legends and people surrounding rare pink river dolphins in "Dance of the Dolphin." More somberly, biologist Marcos Santos's "Lessons Learned from a Dolphin in Brazil" elucidates some of the problems associated with dolphin-human interactions in a more modernized area of South America where legends seem to have been all but lost. Here, Santos attempts to mitigate the problems that arise when a sociable wild dolphin, "Tiao," is abused and then kills one of his abusers in self-defense—the first documentation of a human mortality associated with a wild dolphin.

In Nova Scotia, an example of how one modern-day beluga lives among eastern Canadians is described by Cathy Kinsman in "Luminary." Kinsman's study of this and other solitary belugas illustrates the range of experience in the dolphin-human bond.

The phenomenon of solitary sociable dolphins in European waters is explored in "Solitary, Yet Sociable," by Christina Lockyer and Monica Müller, biologists who have been studying them for decades. In "Life at Monkey Mia," noted cetacean biologist Rachel Smolker takes us to Australia, to one of the oldest and most famous locations where people have been wading into the water to meet and feed wild dolphins for many years. Israeli researcher Oz Goffman tells the moving story of a "Miracle Dolphin" named "Holly" and her calves as they befriend a local Bedouin tribe in the Red Sea in Sinai, Egypt.

Finally, Erich Hoyt—a pivotal figure in the expansion of whale- and dolphin-watching—takes us to Japan and other areas around the world in "Toward a New Ethic for Watching Dolphins and Whales."

JIM NOLLMAN

Talking to Beluga

R iding in a station wagon northeast across Finland through uninterrupted forest of birch and spruce. My driver, Rauno Lauhakangas, informs me we are passing through the lake district, then hands over a roadmap displaying a labyrinth of blue spreading a hundred miles in every direction. Noticing a road sign displaying a moose crossing, I compare the sway of these Finnish hills to the New England of my childhood, only to be interrupted by Rauno. "The name of my country is Suomi, not Finland. The error got started by the Romans, who referred to a northern people called Finns." Then he brightens, "But I like having a fin appendage. I'm just like a whale."

Rauno is a nuclear physicist, affiliated with the Cyclotron in the European Organization for Nuclear Research (CERN) in Switzerland. CERN is where the World Wide Web was invented, and when Rauno first explained the new medium to his son, the boy responded that the Web should be used to save the whales. That answer prompted Rauno to start the Whale-Watching Web (www.physics.helsinki.fi/whale), founded on the economic premise that wherever whale-watching flourishes, whaling inevitably withers. Recent studies have made it clear, for instance, that the growth of ecotourism in Japan is influencing that whale-consuming culture to appreciate living whales more effectively than twenty years of foreign protest ever accomplished.

Rauno is also an avid student of the petroglyphs that adorn so much bedrock throughout Scandinavia and Russian Karelia. The rock carvings were carved by Finno-Ugric tribes who migrated in waves into Europe after

Figure 1

the last Ice Age, including the ancestors of Finns, Hungarians, Estonians, and Saamis. The subject matter includes animals, ships, and depictions of Paleolithic village life so effusive in detail they could have inspired Breughel. The well-known glyph shown in Figure 1 depicts a boat with four adults initiating eight young people into adulthood by teaching them how to harpoon a baleen whale. The strange image in Figure 2, believed to be a snake, is carved on a sweeping slab of granite that slips a half inch below the water line of a large lake. When the wind blows hard, small waves surge across the rock, forming eddies at the raised angles, giving to the snake the distinct impression of movement. Some archaeologists declare this glyph to be the oldest extant example of animated motion.

The most spectacular petroglyph sites are at our trip's destination in Karelia, the Russian republic that shares a long western border with Finland. Well into the twentieth century, Karelia and Finland constituted a unified culture even though Finland spent hundreds of years under Swedish domination, while Karelia has been controlled by Russia since the time of Peter the Great. With the "new" Russia experiencing economic meltdown, the provincial Karelian government is turning increasingly to its prosperous Scandinavian neighbors for aid and codevelopment projects to help reestablish their common heritage. This initiative has gotten Rauno's full attention.

One of the most famous Karelian petroglyph sites, located where the Vyg River empties into the White Sea near the town of Belomorsk, displays a 6,000-year-old image of an ancient Finn and a beluga whale. Scholars believe the glyph is the world's oldest picture of a whale. The fact that belugas still reside nearby has spurred Rauno to develop a program of whale-watching and petroglyph-viewing to ignite Karelian eco-tourism. Scandinavians will surely want to experience their prehistoric roots in Karelian bedrock while catching a glimpse of the last beluga whales in Europe. At Zaitsky Island, just a few kilometers south of the beluga whales' prime summer feeding

Figure 2

ground, is a remarkable site that is the largest labyrinth in the world. Thirty kilometers offshore lie the Kusova Islands, composed of huge slabs of peeling pink granite that seems to glow with an inner light during long summer evenings. Kusova also has prehistoric tombs marked by boulders stacked into eerie human forms, which resemble the sculptural *Inukshuks* of the ancient Dorset culture in Arctic Canada. Nearby Solovetsky Island contains the most famous medieval monastery in Russia, as well as 200 beluga whales residing just offshore of its onion-domed steeples.

Russia was one of the world's most active whaling nations until the 1980s, and many outside observers fear that the current economic pessimism could easily entice the government to start it up again, perhaps focusing on easy-to-find species like belugas. This is precisely the reason Rauno is promoting a whale-watching economy for the White Sea.

I am a quirky addition to Rauno's whale-saving ambition, an American musician and author, best known for playing music with animals. I have worked for the Smithsonian blowing harmonica among howler monkeys in Panama, played flute with wolves, produced radio shows with turkeys, and convened a reggae band to interact with pilot whales off Tenerife. I have even invited Tibetan lamas to recite their prayers through an underwater speaker system in the vicinity of orcas.

A few years back, I created a theatrical performance on the subject of interspecies communication. My Helsinki performance was promoted by Rauno, who publicized the event with a poster displaying the petroglyph in Figure 3, which he had copied from an obscure book about Karelian rock art written in the 1930s by the Soviet archaeologist V. I. Ravdonikas.

Figure 3

Ravdonikas had discovered the glyph covered in lichens along a sweep of bedrock beside the Vyg River.

The image seemed to reflect my own musical experiences with whales, and so I wrote an interpretation of the glyph, which led to an invitation to attend an international petroglyph conference held in the Karelian capital of Petrozavodsk. It was there I learned that belugas still reside near the site where Ravdonikas had made his discovery. When I later presented the same paper at a whale conference in Tokyo, it caught the attention of a Canadian filmmaker who quickly financed an expedition, including Rauno and me, to scout the White Sea whales, the petroglyphs, and the Russian inclination to protect both.

The only previous assessment of this petroglyph was made by the Karelian folklorist Eero Autio, who described the rear figure as "an unidentified sea animal." Actually, the beast's bulbous forehead, up-curved rostrum, blowhole, and lack of a true dorsal fin make for an anatomically precise rendering of a beluga whale. The diagonal line slashing across the image is an ancient crack in the bedrock that clearly gives the animal the appearance of spy-hopping. This is no coincidence, because belugas do spy-hop while observing people on the shore.

The position of the human figure's hands holding oblong objects suggests the performance of a percussion instrument struck like the clavés that are so popular in Latin music. This interpretation draws from the persuasive work by Russian ethnomusicologist Alla Ablova, who has discovered lithophones (essentially stone clavés) at several petroglyph sites in Karelia. Humans playing music is a common motif of traditional art; for instance, the Hopi demigod Kokopelli is always shown as a hunchback playing a flute. However, that this petroglyph figure appears to be playing music with a whale creates an image unique in prehistoric art.

Although the image is unique, the activity is more common than one might suppose. Aboriginal shamans on Australia's north coast have been documented slapping oblong-shaped "dolphin sticks" to attract bottlenose dolphins who respond by corralling fish onto the shore, which are then gathered by the entire tribe. Intriguingly, when asked what quality of the sound caused the dolphins to respond, the shaman Jackson Jacobs answered that the sound was not for the dolphins but to calm his own mind in preparation for telepathic communication. Half a world away, Jacques Cousteau documented the Imragen of Mauritania beating sticks on a beach to prompt dolphins to drive mullet onto a beach.

The figure in the foreground of the petroglyph has been described by Autio as a "humanoid wielding a weapon," ostensibly a harpoon, although why a weapon sprouts from the elbow and points away from the whale has never been explained. Some archaeologists speculate that it is an ancient example of the universal symbol for movement, suggesting a ceremony or dance. In this case, the figure is described not as a hunter, but as a shaman performing a ritual.

The speculation that Karelian shamans may have developed music to engage beluga whales is given a boost from an unlikely source.[1] In A.D. 1074, the German monk Adam of Bremen journeyed through northern Scandinavia on a mission to convert the pagan culture to Christianity. Writing in Latin, Adam observed:

> People of the Northern countries are Christian, except those who migrate along the sea coast near the Polar ice. It is said that these pagans possess magic to communicate with one another when far apart in the world. In addition, they chant powerful incantations in a murmuring voice to persuade great whales to venture close to shore. These tribes know many things which the Bible explains as the talents of wizards.

Adam's use of the Latin phrase *murmure verborum* is translated by the historian of whaling Ole Lindquist as "overtone singing," a vocal technique still performed by Saamis today. The singer simultaneously produces two or more harmonically related notes, which resonate in odd ways and even produce the effect of ventriloquism.

The Solovetsky Island village of Solovky is composed of mostly log-built homes surrounded by potato patches, with few stores, none larger than a single room. The soil freezes to several feet deep each winter, so there is no centralized plumbing, and drinking water is collected by the villagers at pump houses located at key spots around town.

What makes Solovky unique is its Kremlin, an exquisite fourteenth-century monastery that sprawls across five hectares through the center of the village. The complex is surrounded by a dry moat and by high walls constructed of pink granite stones weighing many tons apiece. Inside the gate rises one of Russia's oldest cathedrals, with onion-shaped domes containing huge bells that are rung twice a day. During the Soviet period, Stalin turned the monastery into one of his most infamous gulags, and, touring the complex today, one cannot help but notice walls pockmarked by bullet holes, an effect that leaves a more chilling experience of the Stalinist terror than any historical accounting.

In an obscure book of tales collected from the Karelian oral tradition, Rauno discovered a myth about the origin of the monastery. Two monks, Izootei and Savaatei, journeyed to Solovetsky Island searching for a place beyond worldly distraction. When the sea froze, they called upon a mysterious third companion, a white horse, to pull a sleigh containing the cathedral's huge cornerstone across the ice. But the ice broke, pulling the stone under, and drowning the tethered horse. The dismayed hermits prayed to the archangel Michael for aid. He responded by lifting the stone from the sea bottom, placing it at the alter site, and then restoring the life of the white horse. Because this myth was gathered from villagers living hundreds of miles south of Solovetsky, Rauno speculates that the story may be a fragment of a longer account that was naturally altered over hundreds of years by people who had no inkling of the White Sea, the monastery, or the whales. Rauno's attention focuses on the white horse, drowned and resurrected at the very place beluga whales have resided since the last Ice Age. In other words, the story is a creation myth not only for the monastery but for the local whales as well. This connection becomes irresistible when one realizes that these whales have been listening to the church bells ring twice a day for six centuries.

The Moscow-based Shirshov Institute has established a beluga whale research station a few miles north of the cathedral, where marine biologists have erected a log tower at the tip of a skinny, boulder-strewn glacial esker that all but disappears at high tide. This vantage offers a stunning view of the whales that congregate at the esker's drop-off to exploit a tidal upwelling that attracts schools of herring. The belugas visit the area in groups of five to fifteen animals, usually preceded by a "scout" who signals the rest of the pod if the fish are present and the humans are not acting offensively.

The director of whale research is Vsevelod Bel'kovich, a soft-spoken biologist in his sixties who has been studying belugas for thirty years. He believes the species possesses a language, which he first suspected while witnessing a group of females surround another female giving birth. "I heard them offering her vocal encouragement. When the baby emerged, the group expressed joy to one another. Then they all turned to congratulate the mother."

Beluga whales are the noisiest cetaceans, with a repertoire variously described as the lowing of an ox, bird warbling, cow mooing, grunting, barks, whistles, buzzes, whinnies, chicken clucking, slapping wooden boards together, and the violent shaking of a tin tray. They are also the only whale species to naturally vocalize in air, which they do so often and so melodiously that they were dubbed "sea canaries" by nineteenth-century whalers who marveled at their chirps and chortles. I once recorded belugas in the Canadian High Arctic. Their vocalizations echoed off the pack ice, reminding me of an energetic conversation heard through a thick wall that obscured individual words, although the rise and fall of intonation clearly indicated a language.

Bel'kovich concedes that he does not yet know how to prove his profound hypothesis of beluga language, but he encourages his students to figure out how to "crack the code." He agreed with me that music is an apt tool for exploring the potential for communication with the melodic-whistling beluga whales. Bel'kovich does not hide a concern that human beings are currently killing off the last best chance we may ever have to converse with another species. "ET is alive and well in the White Sea," he declared sincerely. "But without more protection, they will be extinct here within ten years."

The beluga's biological name, *Delphinapterus leucas*, means "dolphin without a fin" because the animal has a thick ridge "fender" running along its back, which profits this animal that spends so much time cruising under ice. Taxonomists place belugas and narwhals into a distinct family because their neck vertebrae are not fused, permitting them to turn their heads with great flexibility. Beluga babies are born charcoal gray and slowly lighten in color over time, becoming pure white by their third year. Newborns are one and a half meters long, whereas a large male attains six meters. Belugas are known to live twenty-five years in the wild.

This whale is also unique for its ability to contract facial muscles. For some years, a beluga whale confined in a tiny pool at a Canadian aquarium spent several hours daily peeking through a porthole at children, mimicking their smiles, frowns, and lifted eyebrows. Although several oceanariums around the world keep belugas in pools, the species is notoriously uncooperative at learning the standard antics like jumping through hoops. Of nine known beluga whales born in captivity, only two have survived even a week. The video of one such birth and death at an American zoo was described as "having the look and feel of a mercy-killing administered by its mother."

Although orcas are the natural predator of belugas, it is human beings who stand overwhelmingly responsible for their decline. Until the Renaissance, belugas (or white whales) were common along the north European coast, feeding in rivers, including the Rhine, the Neva, the Oder, the Thames, and possibly the Seine. All these populations were killed to extinction by the late seventeenth century, mostly to provide heating oil and a supple pure white leather. Today, belugas are threatened or endangered throughout most of their range, yet they are still killed in Canada, Alaska, and Siberia through an "aboriginal dispensation" granted to native hunters, who have relied on beluga blubber (*muktuk*) as a source of vitamin C for thousands of years. Today, no one besides the Koryak of Siberia's Chukchi Peninsula still hunts belugas with nets and harpoons. When the Koryak set off from shore, their wives dress in ceremonial finery and wear crystal charms to soothe the whale's spirit, inviting it to offer its life so the tribe might survive another winter. After a successful hunt, the Koryak perform ancient dances that guide the dead whale to its spirit home. Their song declares that, when the last whale dies, the Koryak die as well.[2]

Besides this single exception, nowhere else is survival a valid reason for native people to kill the species. And because belugas often enter river shallows, they are especially vulnerable to native hunters using rifles. Canadian

Inuit defenders of the hunt point to the faceless marketplace to argue that killing whales helps them maintain traditions that link their people to the sea. But in truth, the hunters who argue the loudest often show little regard for tradition, because ironically they are not so much lovers of heritage as lovers of guns, fast boats, and the adrenaline rush that comes from killing large animals. This has led some critics, including the distinguished American environmentalist Edward Abbey, to decry natives who kill endangered species under a special "aboriginal dispensation" as "not hunters but merely gunners."[3]

Commercial whalers took belugas until the 1980s. Although individual animals are relatively small when compared to a baleen whale, the species is covered in a thick layer of blubber that renders a reasonable amount of precious oil. Because belugas congregate in large social groups called pods, many animals could be killed at one time. Today, oceanic pollution poses more of a threat to belugas than the revitalization of commercial whaling. Arctic waters act as a sink for the flow of pollution in the world's oceans. As a geographical group, arctic polar bears, seals, and several whale species, as well as the Inuit who consume these creatures, possess dangerous concentrations of heavy metals, PCBs, and radioactive contaminants. Just north of the White Sea lies the Soviet military's notorious nuclear dumping ground where power plants and submarines have been jettisoned.

During our expedition, Rauno and I spent much time on our hands and knees creating several rubbings of the beluga shaman petroglyph (Figure 4), and we made the profound discovery that the archaeologist Ravdonikas left the world a wildly inaccurate portrait of the famous petroglyph. The best explanation is that he probably sat in front of the poorly preserved image and simply drew what appeared to his eyes. By contrast, a rubbing exposes every faint incision and pockmark.

Ravdonikas completely missed a star-shaped object above and behind the whale. Nor did he see the circle to the far right that ostensibly denotes the sun. Add in the crescent moon, and the three objects presumably define a triangulated positioning of celestial bodies, which suggests a method still used today to determine date and time. If so, the glyph would have to be regarded as the world's oldest performance poster, informing the tribe of the time and date for their shaman to perform an annual whale communication ceremony. Or, time and date might point to the past rather than the future, thus commemorating a historical event when a shaman and a whale sang together.

Figure 4

Our rubbing also sheds new light on Eero Autio's purported harpoon. There is no arrow head on the end; it is, rather, a tongue-shaped appendage that originates not at the elbow but continues under the arm to disappear inside the shaman's mouth. The Russian petroglyph expert at the Hermitage, Avrim Stolyar, has pointed out that many images rely on exaggerated features to emphasize the capabilities of those depicted. Huge feet describe a fast runner, big hands depict a talented hunter, a long snaky penis describes a man with many children. The beluga shaman glyph thus portrays a human with a very long tongue, the attribute of a talented speaker or singer. Intriguingly, the beluga is sporting its own "word balloon"—a dynamically articulated spout emanating from its blowhole. It thus seems likely that 6,000 years ago a shaman and a whale were captured by a Karelian artist while they communicated *(murmure verborum)* to one another.

The shaman has a pronounced swell at its breast, which indicates that the figure is a woman. A small crescent moon—the universal symbol for the feminine—floats in the air adjacent to the figure's breast. And this woman has been caught in the surreal pose of giving birth to a whale, which, as strange as it seems, is a common mythical motif of totemic people all across the Arctic.

Whenever the opportunity arose on Solovetsky Island, Rauno and I boarded a dinghy to motor midway between the whales' feeding and resting areas. We anchored 200 meters from shore and 100 meters from the whales' normal "path." The water was so clear we could easily discern our anchor lying on the green rocks thirty feet below.

We spent several hours a day over a week's time trying to engage the belugas with live music. Every session started the same way, by striking a brass Tibetan meditation bell for at least ninety minutes. This sound became our "signature," piercing but calming, producing distinct beats as the tone emanated across the open water. Bio-acousticians now know that several different species of toothed whales communicate by employing such signature sounds in their own communication. Since every whale in a group has its own unique signature, beginning a series of calls with a signature establishes the identity of the caller. The first time we tried the bell, it attracted the attention of an advance scout, who swam so close to the boat we could have touched it if we had been less respectful. The sinuous white whale tinted green by the water slowly swam circles around us, then turned its pliable neck to stare straight into my eyes.

Playing the bell evolved into a routine, consisting of a minute-long repeated pattern of sound followed by thirty seconds of silence. A few hours into this rhythm, five whales arrived to chirp and neigh but *only* during the periods when the bell was silent. They repeated this synchrony enough times to dispel any disagreement that the timing was just a coincidence.

Few people would deny that music conveys emotion across culture despite the language barrier. Science-fiction films make the case that music is the obvious choice to establish communication with extraterrestrials, should the opportunity arise. My experience recommends that this same emotional transparency translates across species as well as culture. I do not argue that the whales understood a specific "meaning" to the bell, but that the signature imbued our effort with the sincerity we felt.

I eventually switched to electric guitar, favoring a simple chord progression played in a reggae rhythm, and transmitted underwater at about the same volume as our ten-horsepower engine. Reggae creates distinct "holes" of silence in the syncopated rhythm, which offers a metrical invitation for the whales to fill. But nothing happened.

It was a different story the next day. Although the sun was shining when we anchored, the air smelled of rain. Within an hour, two distinct thunderstorms darkened opposite vectors of the sky, rumbling and flashing lightning as I stowed the bell and plugged in the guitar. Within fifteen minutes, five whales drew within 100 meters of the boat, vocalizing notes within my pentatonic scale. The scout left this group, swam beside the underwater speaker, then turned to rejoin the pod. Within another minute, whales began vocalizing within the silent spaces of the reggae progression. The session ended quickly when a bolt of lightning struck the sea too close for us to ignore.

Rauno had his own instrument, a "phrase sampler" gadget that let him record and archive short bursts of beluga conversation, modulate the sounds in various ways, then push a button to send the modulated call back into the water. By the fourth day, he had collected several different calls in what we termed his beluga "dictionary." For no special reason, he treated a single call as a single word, although it seemed very long odds indeed that pressing various buttons in succession would conceivably speak a beluga sentence.

The potential of the sampler has deflected our future research in a more hi-tech, lingual direction. We are altering pattern recognition software to help ascertain, for example, which individual whales make which sounds. Ascribing calls to specific individuals is a first step that could one day led to the actual meaning of those words, and similar techniques have already led to profound results unraveling lingual aspects of sperm whale calls off Tenerife and bottlenose dolphin communication near Scotland. We eventually hope to develop a kind of acoustic computer game for belugas to control by their voices.

If belugas have a language, will it irrevocably alter our species' self-important role as the only planetary intelligence? In its own way, the revelation seems the equal of Copernicus proving the Earth revolves around the sun. As we grapple with such profundities, how strange that the interspecies bond Rauno and I encouraged sometimes seemed effortless to establish. It was too easy, as if the belugas had been waiting patiently for 6,000 years for us to re-learn how to greet them. After a week spent among them, they offered us enough signs that they were as interested in exploring the acoustic interface as we were. Who dares speculate how the relationship might evolve over an entire summer spent in their presence? With enough time, perseverance, and sponsorship, Rauno and I will definitely return to this place again and again.

NOTES

1. Brought to my attention by the Icelandic historian of whaling Ole Lindquist. The translation from the Latin is a mix of my own and Lindquist's.

2. Victor Perera, "Of Whales and Men," in *The Nature of Nature*, ed. Billy Shore (New York: Harcourt Brace, 1995): 147.

3. Edward Abbey, *The Journey Home* (New York: E. P. Dutton, 1977): 52.

At Home with the Dolphins

Being adrift on an inflatable boat, not too far from the coast, and surrounded by dolphins is one of the places I would call *home*. When the animals, the sea, the boat, and myself are framed in a single painting, then there seems to be nowhere else to go, and everything feels just fine. To me, this is one of the joys of being a cetacean researcher.

If I were an artist, I would maybe record dolphin sounds and, late at night, improvise on the whistle scales with my jazz guitar. I would protray them in the splashing water as they leap or write dolphin haikus on waterproof paper. But I happened to become a marine biologist, and, as a scientist, I choose to interact with these animals on a different level, through some sort of filter that prevents thoughts from ranging dangerously wide.

When I began, I knew little about dolphins. I started working with these animals because I thought that they were special and smart. And indeed they are. But I do not see them as big-brained geniuses or enlightened philosophers. I certainly do not see them as beings that stand apart from the rest of the living world. Instead I look at them as inseparable parts of a complex net of interrelated beings, and the more I know about them, the more I appreciate the marine environment as a whole. I count dolphins among the many, multifaceted life forms that have emerged from a multi-million-year evolutionary process. They are incredibly well adapted creatures, living expressions of whatever beautiful and precious lies in the oceans.

By working with the dolphins, I started to care about them. And when you start caring about animals and see how each one is important and fragile, when you realize how every living being is interconnected with all other living beings, then you start caring about *all* animals. Dolphin, but also fish, squid, bird, turtle, krill. And then slug, elephant, mouse, myriapod, snake, frog, spider. Dolphins are very good at teaching this lesson of interconnectedness, because they are top predators that depend so much on all the other marine organisms. Take away the krill, contaminate the plankton, reduce fish stocks through intensive trawling, catch the sharks, and dolphins will be affected.

When I observe dolphins in the eastern Ionian Sea or in the northern Adriatic Sea, where we have been recording behavior for some 2,000 hours over the past decade, I can see the effects that human behavior has had on them. Bottlenose dolphins are spending too much of their time looking for food. My colleagues take photographs of all the dolphins encountered, and they find out that a great deal of them are underfed, while remotely collected skin and blubber samples indicate heavy contaminant loads in their bodies. These are not the only discouraging findings. If one goes through the literature and checks the historical data for the central Mediterranean Sea, it is startling how many things have changed in just a half-century. Dolphins' food prey has been dramatically reduced, and marine food webs are heavily contaminated by PCBs and other man-made xenobiotics. Bottlenose dolphins are declining, and the short-beaked common dolphin is no longer "common." It is a sad legacy.

In response to this, my colleagues at the Tethys Research Institute[1] and I have provided scientific data to be used for proper resource management, and we have striven to improve public awareness of the problems faced by Mediterranean whales and dolphins, with the goal of preventing these animals from disappearing. When I myself started to work on dolphins, to tell the truth, I did not aim at "saving" them. What I really wanted to do was get a little closer to them, being fascinated by their apparent beauty and perfection. But the more I have learned about dolphins, the more I have been impressed by the human-related problems they face. Perhaps because of this conflict I did not think that close interactions with humans could benefit wild dolphins.

But interactions do occur. We may be doing photo-identification[2] and the dolphins may feel like riding our bow wave. They position themselves in front of our boat and wait for it to speed up. But we do not want to affect their natural behavior, and of course having dolphins bow-riding does not enable us to take good photographs of their dorsal fins. So the dolphins wait impatiently, gazing at us with their soulful eyes. Eventually I switch from idle to forward at very low speed, and the dolphins look like they are enjoying themselves, even if the bow wave we create is very small. The dolphins take pleasure in it anyway; for them, even the simplest events may become a great attraction. A fisherman's boat passes near us, and the dolphins rush to surf "waves" that are only a few inches high.

Dolphins are naturally playful creatures, but mostly they are too busy to have fun. Often we observe them searching for prey and performing long dives and ventilation cycles. In these situations, they do not feel like interacting at all, and it is clear that they do not even want us to approach too closely. This makes photo-identification difficult, but we respect the dolphins' behavior. We try to take photographs from a distance and follow them until a change of "mood" occurs. Occasionally we may stay with them for several hours, until the sun goes down. By spending so much time with the dolphins, we have a chance to register their many activities, from serious food-gathering to enthusiastic play.

We can learn more about dolphin psychology by watching "lone sociables," individuals that choose to live in close proximity to humans. One such dolphin is Filippo, an adult male bottlenose dolphin that in 1998 settled in the port of Manfredonia, in southern Italy. My student Giovanna Barbieri, who moved to Manfredonia to study Filippo for her biology thesis, reported:

> He spends most of his time floating beside the same moored boat, at times for up to six consecutive hours, then he "wakes up" and goes playing with boats that are leaving the port. He follows them and, no matter how fast the boat goes, he swims with his rostrum a few inches from the propeller. Sometimes he masturbates with the rubber parts of an inflatable (he is a very libidinous dolphin), or allows the people on board to pet him. After making a few longer dives to catch food, he returns to the port to rest beside his favorite moored boat.

When people come to visit Filippo, they do not believe that what they are seeing is really a dolphin. Giovanna points her finger at a brownish buoy—the dolphin's melon and foresection—that floats kind of motionless near the harbor master's boat, in a polluted port where no sentient being would even dare to swim. "That's Filippo!" she declares.

Filippo began to interact with humans, and particularly with professional diver Pasquale Sdanga, in 1997. At that time he was kind of shy and wary of humans. But once the first physical contact was made—a gentle touch by the diver's hand—his behavior toward humans changed dramatically. Although he maintained a special relationship with Sdanga, he became increasingly rough and even aggressive with bathers. Filippo clearly likes to interact with people but always claims a dominant role in the interaction. He enjoys games that can upset even the most skilled and patient swimmer; Filippo is a powerful animal that can easily overwhelm his human playmates. If a swimmer does not cooperate satisfactorily, Filippo may get nervous and become even more demanding, at times kicking with his rostrum or even biting. Although his behavior may be fair whenever he is in a "good" mood, he is rather unpredictable, and our suggestion to managers and authorities has always been to prevent people from swimming with him. A number of minor incidents have occurred, wounding both swimmers and dolphin: Filippo is still a wild animal, not a pet.

People have so many expectations and beliefs about dolphins that it may be hard to see them as wild animals and treat them accordingly. Popular television and oceanarium shows have taught us that these are ever-playful animals, always happy to interact with humans. This has affected our capacity to relate with them, sometimes with dramatic results. On August 28, 2000, the local press reported news of a dolphin group enjoying a "summer party" in the coastal waters off the island of Ischia—a beautiful and famed tourist destination west of the Italian peninsula. My friends and colleagues Barbara Mussi and Angelo Miragliuolo, however, recall a less than idyllic scenario.

Barbara and Angelo, who started a long-term study on the rich cetacean fauna of the archipelago in 1991, had a chance to participate in this dolphin "party." "Everything started when a group of approximately twenty Risso's dolphins[3] was sighted by pleasure fishermen one mile offshore," Barbara recalled. "Some of the many pleasure boats that crowd the area in the summer months started to approach the dolphins. The number of boats kept increasing, and by midday the harbor master's officers called Angelo and me to get expert advice on how to prevent harm to the animals."

By means of the harbor master's inflatable, the group was located around 1 P.M. By the time the authorities arrived, the dolphins had been surrounded by some 100 speedboats, all with their engines turned on. Nearby, another 400 boats were anchored. To the observers' astonishment, dozens of boats were "playing" with the animals by heading toward them at high speed every time they surfaced, changing route suddenly in order to take pictures, and attempting to get closer and closer to the terrified dolphins. The animals were herded ever closer to the resort harbor, ending up in water that was only ten feet deep. Barbara reported that all the dolphins were showing clear signs of distress and seemed unable to orient themselves. They swam erratically at high speed, at times colliding with one another. Of at least three calves observed in the group, one was seen spinning and swimming in circles apart from the others. Angelo tried to prevent a mass stranding by placing the boat between the animals and the beach, but the dolphins responded by splitting into two tight subgroups, one of which included the calves.

In the meantime, none of the pleasure boaters appeared to realize what was going on: they were simply excited to see the animals at close quarters. Through a two-hour effort, the researchers and officers finally managed to clear an area of pleasure boats. As soon as the "opening" was wide enough, the animals formed a single tight group and slowly started heading offshore. The Risso's dolphins' "summer party" eventually came to an end as the animals moved into safer waters, "escorted" at a distance by their exhausted rescuers.

Occurrences of this kind of human-dolphin interaction are becoming routine worldwide. Many seafarers still have to learn how to relate to cetaceans. Not far from Ischia, for instance, the waters of the International Ligurian Sea Cetacean Sanctuary[4] are being increasingly affected by heavy boat traffic that can have a negative impact on cetacean populations by posing threats including ship collisions, high ambient noise, and disturbance from unregulated whale-watching activities.

Quietly watching cetaceans in the wild while they busy themselves with their daily activities offers an unforgettable experience, as well as a chance to learn. The basic lesson that I think I have learned by spending time with dolphins in their natural environment is that they are an integral part of it. I do not like watching dolphins in a tiled pool. Dolphins in captivity are often supposed to have an educational function. But seeing a dolphin in a pool gives little idea of the connection between the animal—or any organism—and its natural habitat, not to mention the right of an animal to live in its own natural setting. The bond that ties an animal to its environment is not something that should be overlooked, and a dolphin out of the sea and out of its complex, fluid society can hardly represent its species. The key to environmental education lies in teaching people how to understand and protect ecosystems, rather than single animals. What kind of dolphin is a dolphin set apart from the sea?

By working at sea, one learns that the animals there are part of a beautiful, complex system. They do not always perform the "tricks" we like to watch. They may not even feel like being approached. A different way of relating to wildlife arises from days spent just waiting for the wind to stop, from hours of floating adrift, and from scanning countless miles of sea surface for dorsal fins. In the end, it does not matter if the dolphins leap or dive, if they come close or shy away, because you know that they are there—that they are part of this sea—and you can see signs of their presence in schools of jumping fish, in the shearwater passing by, in the dancing jellyfish. It is still a place worth living in—the sea still feels like home.

NOTES

1. The Tethys Research Institute (www.tethys.org), founded in 1986, is a nonprofit, nongovernmental organization dedicated to the preservation of the marine environment. It focuses on marine animals and particularly on cetaceans inhabiting the Mediterranean Sea and aims at protecting its biodiversity by promoting the adoption of a precautionary approach for the management of natural resources. It is based in Milan and Venice, Italy.

2. A technique that consists of photographing body portions carrying long-term natural marks suitable for individual identification. With dolphins, dorsal fin shape and marks on its trailing edge, such as nicks and notches, are often used to recognize individuals through the matching of photographs taken in the field.

3. This dolphin species is relatively common in the Mediterranean Sea, particularly on steep continental slopes. It is seasonally present off Ischia, possibly attracted by increased prey availability around productive submarine canyons.

4. Following a proposal made by the Tethys Research Institute in 1990, and a decade of intense lobbying coupled with public awareness and education, on November 25, 1999, Italy, France, and the Monaco Principality signed the final agreement for the creation of this sanctuary. Aimed at preserving cetacean populations and promoting a sustainable use of marine resources, the approximately 260,000-square-mile sanctuary is the largest protected area in the Mediterranean Sea.

Dance of the Dolphin

AUTHOR'S NOTE: *This research began, for me, in Bangladesh, while looking for man-eating tigers. Exploring the muddy rivers of Sundarbans, the largest mangrove swamp on earth, I was researching my book* Spell of the Tiger, *on the aquatic tigers who swim out after fishermen's boats and devour some 300 people a year. I saw no tigers for weeks, for the flooded forest there concealed them. But one day the muddy waters parted for an instant, and revealed, to my astonishment, the curves of three pale pink fins. They were Gangetic dolphins—one of the world's five species of strange, little-known, river-dwelling whales. And although I only saw them for an instant, their image continued to haunt me.*

Years later, at a marine mammal conference, I met a man who told me why. "Pink dolphins capture souls," he said. Half a world away from Sundarbans, in the Amazon, he told me, lives a different species of pink dolphin.

His photographs intrigued me: the pink Amazon river dolphin looks nothing like the athletic marine dolphins who leap in aquaria and sport in front of boats in the ocean. A large animal weighing perhaps 400 pounds, the pink Amazon river dolphin has a curious bulbous forehead, long tubular snout, and wing-like foreflippers with five fingers inside, like ours. They are the world's most primitive whales. And unlike the Gangetic dolphins I had seen in Bangladesh, these Amazon dolphins are bold and abundant, I learned. But they are equally mysterious.

A number of people at the conference were trying to study them, with little success. Few claimed to know the number of dolphins in their study area, or to understand the social structure of groups. They did not even know for sure why these dolphins are pink: some said these Amazon dolphins glow pink with exertion; some said only the biggest and oldest animals are pink, and the others are gray or whitish; and most researchers agreed that, even after years of study, it was almost impossible to recognize individuals on sight.

To local river people, this confusion is no surprise, my informant told me. The people say these dolphins are shape-shifters. They say they can take on human form, rise from the water, and show up at dances to seduce men and women. They can claim your soul and carry you away to the enchanted city beneath the water, the Encante—a place so beautiful, no one ever wants to leave. In fact, these river dolphins do inhabit an enchanted world: the Amazon.

I had always longed to go there, but its scale was overwhelming. Discharging fourteen times the flow of the Mississippi, source of half the world's river water, the Amazon's jungle is the size of the face of the full moon, and its diversity staggering. There are fish here who nest in the trees. Fish who sing with electromagnetic signals. Ants who garden, creating a substrate of leaves on which they grow a fungus found nowhere else in nature, their only food. Here live birds with claws on their wings, trees with spines that sting, fierce otters six feet long, spiders the size of your hand, lily pads big as throw rugs. Surely I could never fathom this impossibly huge, strange, rich land without a guide. But now I had found one.

Dolphins have guided humans for millennia. Legends older than the Minoans tell us dolphins led the ancients to the center of the world. Even the Greek Sea God, Poseidon, turned to a dolphin to help him find his lost bride. So in the Amazon, what better guide could I choose than the pink dolphin? I knew I had found my next book. I had read that these dolphins were thought to undertake long-distance, seasonal migrations, and on this premise I proposed my journey.

With my photographer and friend, Dianne Taylor-Snow, I would follow the dolphins on their migration. This was not, in fact, what happened.

I met them first at the famous Meeting of the Waters near Manaus, Brazil, where the dark waters of the Rio Negro meet, but do not mix, with the creamy, silt-laden waters of the Rio Solimoes. Here a Brazilian researcher, Vera da Silva, had begun her study of these enigmatic animals, outfitting them with radio telemetry at a later research site I would also visit, farther south, at the huge flooded forest of Mamirauá. Her studies proved they were not migratory after all—research she had not yet published when I submitted the proposal for my book. But, still, I would follow these dolphins in ways I never could have imagined: I would follow them on a voyage back into time, and into a world where the plants talk, the trees walk, and the whales come out of the water and dance with humans on the land.

It was not until my last expedition that I would finally see the dolphins whole, in clear water—and come to know them as individuals, as a biological species, and as agents of ancient, magical powers that connect the people and the waters with the spirits and the land.

"Our legend is that there was a very pretty girl in one family. Here, you see, we used to hold a ball—the Rose Ball—and before the dance, the girl came to bathe in the river. She didn't notice she was being watched by a boto. But from that time on, the boto wanted her."

Through our translator, Isabelle, Necca is telling us the story she learned from her grandparents. Necca's full name is Ludinelda Marino Gonçalvez, and she is forty and wiry, with the high cheekbones of a Bourari Indian, a heritage of which she is proud.

Even though we hear her words in translation, as we sit beneath the thatched roof of the cabana where she sells fried fish and soft drinks at the edge of Lago Verde, at Alter do Chão on the Rio Tapajós in Brazil, we can see Necca is a master storyteller. She reveals her story slowly, slyly. As she speaks, her eyes slide from corner to corner, as if spotting some detail there to remind her of the next turn of events.

"The night of the ball," Necca says, "the beautiful girl arrived with her boyfriend. But the boto [the Brazilian name for the pink dolphin, believed to be a shape-shifter who can emerge from the river in human form to enchant local people] was the handsomest man there. She looked at him and fell in love—and he with her. They were blind for one another."

Oh yes, Dianne and I agreed with a glance. We knew that feeling: when love floods the senses, jams your sonar, blinds you to all else. Lightning might crash around you, eels and piranhas nibble at your toes, and you don't care, because only One Thing matters—that longing which has overtaken your soul. We humans were made for that sweet, sweeping sickness. But what if you have fallen blindly, impossibly in love with a dolphin?

Necca continued: "One man realized that this was a boto. He chased him away. But the girl loved him. She loved him madly. At the next ball, she came and looked for him. There he was, and they danced all night in each other's arms, and then they walked on the beach.

"They lay down in the sand, and there they made love. They did *everything* with each other," Necca said, inviting us with her eyebrows to imagine the details. But after the lovers fell back into the sand exhausted, he suddenly leapt up and disappeared into the water.

"She was heartsick. She went to the village *curandeiro*, hoping he could help her find her lover. The *curandeiro* asked help from the Mother of the Lake. The Mother agreed to ask the Moon to call him back. For the girl was now pregnant. And the *curandeiro* saw that the father of the child was a boto.

"At the next full Moon was the next ball. The Moon called to the boto, and he came to the ball to meet his love. And then the girl told him she was carrying his child. Now he was forced to explain: though he loved her and longed to be with her, he could see her only at the balls, for he could change into a man only on those nights.

"It was months before the next ball. When they met again, she brought him his son. Every ball thereafter he came to see her, to dance with her and see their son."

Necca choreographed a dance to tell the story. Our boatman, Braulio, had told us this earlier: the dance is performed each year, the last week in July, at a festival at the water's edge in Alter do Chão. Necca has danced in it, and now her beautiful twenty-four-year-old daughter, Keila, performs; when Keila's young daughter is old enough, she will dance, too.

"Many people, old people, believe this story," Necca told us. "Many girls today come to the balls hoping to meet a boto! For they are the handsomest men there. And when people come to see the dance, dolphins often come near the beach, as if they knew the dance was about them."

As rain was our companion on earlier journeys, now our companion is fire. We had welcomed the rain, flooding the world with its brimming abundance; but though we dread the fires, we cannot escape them. They are set by corporations, to clear forested lands to plant with corn or farm cattle—farms that last only a few years, leaving the land barren. Some days, on the horizon, there are four or five columns of smoke; some days the air is leaden with it, pressing the river flat as quicksilver. One night, fire had nearly come into Alter do Chão itself, so close we could feel its gnawing heat. Always, somewhere, there is fire or smoke, insistent reminders of the greed consuming the world.

As we head to Curucu this morning, where we know we will find the dolphins, we see three big fires to the west, and the air is hazy. Only in the river can I find respite from the burning. As Dianne unpacks her photo gear, I swim out to meet the first fins. I breaststroke slowly and they come to me: first, the mother and baby; one very large pink adult; two big grays.

I am surrounded, but I do not know this; Dianne, taking photos from the top of the *Gigante*, later tells me there were dolphins all around me. Two pinkish tucuxis, the little marine dolphins of a separate species, joined the group, one with a slash across the dorsal. For three-quarters of an hour, there was not one minute that I could not see a dolphin at eye level. As they surfaced, it seemed I could feel their glance. Or perhaps it was their sonar. Later, I would learn that people who work with marine dolphins say they feel the animals "sounding" them, throwing out trains of ultrasonic clicks created by moving muscles in the melon, and waiting for their echo to return. Their sonar retrieves a three-dimensional soundscape, a sonogram, of what lies ahead of them; the boto, with its flexible neck, can turn the head in a wide arc and obtain an exceptionally broad sounding. One researcher, Bill Langbauer at the Pittsburgh Zoo, told me when the sound waves hit him, if feels like humming with your teeth clenched.

But that was inside an aquarium tank. I swim inside of living water, and always I feel the water humming, charged with life like the blood in my veins. I cannot feel the botos sound me, but surely they are doing so. They can see through me, as God does. They do not touch me, but their soundings penetrate my flesh. They know my stomach is full and my womb is empty; they can see the faulty mitral valve in my heart. And yet, even possessed of this astonishing sonar, they still pull their sleek faces out of the water to look at my face. Why would my face be important to them? They can recognize me by the other nine-tenths of my body beneath the water. Yet they look into my face again and again. They must know we humans wear our souls on our faces. Perhaps, to them, too, a meeting is more profound when it is face-to-face.

And what do I know of them? I only know the sex of one—the mother with her baby. I do not know how long they have lived. I do not know how far they travel. I do not know if this group always stays together, or only comes together here to investigate the pale, terrestrial stranger who hangs in the column of water before them.

But somehow I am not frustrated that I will never learn the answers; this is not, I now know, what they have to teach me. Already, their kind had shown me so much: the botos brought me to Manaus, the impossible Paris in the Amazon; the botos drew me to the Meeting of the Waters. The botos drew me to the Tamshiyacu-Tahuayo Community Reserve in Peru, and to Mamiraá, the world's largest flooded forest reserve in Brazil—and now, to the clear waters here. Throughout these four journeys in the Amazon, I had followed them, though not in the way I had originally planned. For the verb "to follow" carries many meanings, most of which I had not been aware of when I had decided to follow them. To proceed behind them, to go after them in pursuit, as I had envisioned when I had hoped to trace their migration, was merely one way of following. Other meanings are more subtle and profound: to be guided by; to comply with; to watch or observe closely; to accept the guidance, command, or leadership of; to come after in time; to grasp the meaning or logic of. And so, without chasing them along a migration, without a single hit on my borrowed radio telemetry, I had followed them nonetheless. In Tamshiyacu-Tahuayo, I had followed them to the spirit realm where shamans commune with the powers of the plants and visit the Encante; with Gary Galbreath, a paleontologist I met there, I had followed them back through time. At Mamiraá, they had taken me to the heart of the Amazon's modern conservation dilemma. And now, I followed them still, hanging in the water before them, ready to receive their gifts and their guidance.

I always save the moment I enter the water for them. Dianne and I wait at the edge of the river, and I do not swim until the dolphins come. I spot one fin, close, and give my body to the water.

Dianne stays at the edge of the river, or sometimes on the roof of the *Gigante*, photographing. She is a strong swimmer, but her father was a sea captain; those who know water most intimately, I have found, tend to more wisely respect its dangers. But I lose myself utterly in the water, like a soul leaving a body.

Now that we have come here daily for a week, we recognize seven individuals: there are two medium-sized grays, the bowler-hatted businessmen. In fact, Dianne knows an Englishman who owns an actual bowler hat, and we named one of the dolphins after him—David. The other we named Gary, after my paleontologist friend. One very large gray dolphin we name after Vera, the dolphin researcher whose radio telemetry we had borrowed. A very large pink one we call Valentino. Another gray with pink lips, and we name him that—Pink Lips. And occasionally we spot a mother and baby. It is nearing Christmastime, so we call them Mary and Jesus.

Pink fins, gray bellies, the bulbous soft heads; I so wish I could touch them. I can feel the currents their bodies make as they slip through the water. One day, I look down into the water beneath me and see a large gray form swim under my feet. All but immersed in these clear blue waters, I feel as if embraced by paradise.

In Curucu, we had found an ideal site. We could imagine no situation better. But we still felt we should visit the dolphins of the Arapiuns, as people had urged us to do.

At the Mercado 2000 in Santarém, we had met a fisherman with thinning hair and strong hands named Valdomiro Ribero who had spent a month on the Arapiuns when he was fifteen. He would never forget it. "There were many, many dolphins there," he had told us. One night when he had slept on the beach, he had spotted two strange-looking girls, perhaps eleven and thirteen years old. Their skin was very white, he said, but had reddish spots. And they had very little hair. These girls, he had been told, were the boto's daughters.

Certainly, there had to have been a reason that several natural history film crews had bypassed our lovely Alter do Chão and made the day-long journey upriver to the Arapiuns to the west. Perhaps, Dianne and I thought, the waters were even clearer there, the dolphins more easily seen.

We found making arrangements for the journey difficult. Not every boatman will risk it. Braulio told us his *Gigante* could not negotiate the current. The portion of the Tapajós leading to the Arapiuns is difficult to navigate—there are dry spots and rapids—and besides, he said, the place is full of mosquitoes.

It took us five days to arrange for a boat. Sadly, Isabelle was leaving us, to visit Brazilian friends in Belo Horizonte, but she helped us get the trip organized before she had to depart. Socorro knew a boatman, Gilberto Pimentel, who agreed to consider the trip. Necca's beautiful daughter, Keila, speaks some English, and she agreed to help translate our negotiations for the boat. Grateful for Keila's and Socorro's help, we invited them to ac-company us. To our surprise and delight, they accepted.

Through Keila, Gilberto explained that his boat was a strong one. It is named *Boanares*, after his father, who accompanies him on his trips. "But even with a strong boat, the trip can be dangerous," he warned. "Many boats break up when the water is angry, and there are many shifting sandbars. The river is so wide that you cannot swim to shore." We should plan on bring-ing supplies in case we got stranded, he told us. In any event, we would need to spend one night on the water, where, he confirmed, there would be many mosquitoes.

We left at four in the morning, when the wind was very still. But after an hour of travel, the waves grew so rough that the little table in the cabin fell over, and all of us but me were thrown from the hammocks we had strung up in the boat's central cabin. We sat on the floor. We stopped for an hour at Ponto des Pedres—Places of the Stones—to wait for the water to calm down.

Another hour's travel upriver, at about fifteen miles per hour, and we reached a spot called Icuxi. Here, Gilberto strung up his seventy-foot-long net, with plastic Coke bottles for floats and a brick of red sandstone for weight. Once the net had caught some fish, he explained through Keila, the dolphins would come. So we waited. We lay in our hammocks. We walked along the beach. We waded in the water. But, although the net caught sev-eral fish, no dolphins came.

We continued on. We passed more white sand beaches. We passed in-numerable fires. Finally, toward sunset, we reached the Rio Arapiuns. Its blue waters were calm. No fewer than nine fires burned on the western hori-zon. To our surprise, the Arapiuns looked exactly like the Tapajós.

As night rose, we strung up our hammocks again, and Dianne and I fitted ours with mosquito nets. The mosquitoes were as voracious as predicted. To our dismay, we discovered that Socorro and Keila had not brought net-ting, but our friends did not complain; clouds of Portuguese were rising from their hammocks in an animated discussion. I imagined their Portuguese somehow repelled the mosquitoes, like a column of smoke from a mosquito coil.

I listened, and to my surprise, even with my rudimentary grasp of the language, I found I could understand their conversation:

"Do you think a boto will come?"

"Will he be enough for four women?"

"And we'll need one for Gilberto! And for his father!"

"Oh, we will not sleep tonight! We will wait for the boto to come!"

"If only we had *festa* music to attract him!"

Rocking in my hammock, I could feel the waves rising beneath me like a lover.

We woke at five and headed to a place called Jari. There were many shallow channels there, and good fishing—for both people and botos, Gilberto said. With Keila, Gilberto took a rowboat out to speak with the fishermen, to see if they had seen any dolphins.

Yes, they said—right over here! They pointed to a channel separating two grassy shores where skinny cattle were grazing.

And sure enough, we spotted three botos: a large pink adult, a large gray adult—and a small, grayish pink baby.

Quickly, Gilberto and his father set the net across the channel, confining the botos to the shallow waters where we could best see them. The water is only waist deep and somewhat muddy. Dianne quickly loads her camera. But in the channel, something has gone wrong. Something is caught in the net—the baby boto! The bright pink adult starts to thrash in the water, showing her flippers, obviously distressed. The baby might drown! I jump into the water—without thinking to remove my sneakers—and immediately sink into the mud up to my ankles. Gilberto and his father jump in, too, wisely barefoot, and rush to the net. I stagger to join them.

The baby is squealing in terror, and with every effort to escape, entangles itself further. Gilberto's father, Boanares, reaches it first. I try to hold the baby still and keep the head above water while Boanares disentangles the flippers and tail. He does not want to tear the net. A seventy-foot net takes more than a week to make by hand, and in a store costs $150. "Cut the net! Cut the net! I'll pay for a new net!" I bellow, but of course Boanares does not understand me because I have forgotten all my Portuguese.

The baby screams in my arms. Its melon heaves with the force of its terror, its blowhole opening and closing, howling like a tiny mouth. The infant is nearly four feet long, and shockingly strong, one big muscle tensed in panic. The best I can do is hold it still, try to calm it down. I slowly stroke its skin, running one hand under the belly, holding its side against my pounding heart. The skin feels like a boiled egg, and is now flushing pink with exertion. I know now the baby will not drown, but I fear the net will cut the delicate skin. I look into its pearly eye. To the terrified infant, I am a lightning storm, a caiman, a pack of piranhas, an evil spaceship. The baby has no way to understand that we are trying to help.

Finally, Boanares frees it from the net. I hold the baby for a moment longer, at once a heartbeat and forever, my clasp around its body a plea and a prayer. And when I release the boto, I let go my heart from my throat, and feel the water surround my empty arms like forgiveness.

But now the mother is caught in the net! Dianne hands her camera to Keila and gets in the water with us to help. The mother is much calmer than her baby. Dianne wants to hold her, and the mother doesn't struggle in her arms. I almost wonder whether, in the manner of mother birds feigning a broken wing, the mother had embedded herself in the net to distract us.

Now the dolphins swim free. The dark one leaps, as if in triumph. Gilberto and his father remove the net. We all adjourn to the orange and blue *Boanares.*

Everyone is exhausted. Socorro and Keila had been terrified that the two adult dolphins would attack us while we tried to disentangle the baby. Gilberto and his father were afraid that somehow the Brazilian environmental police, IBAMA, would find us with a dolphin in our net and arrest us all. And Dianne, though seeming cool and poised, was worried, too; she was afraid that she would miss the shot. But she didn't.

I collapse with emotion—the thrill of holding the baby, the terror that it might be injured, the guilt that had a boto been hurt, it would have been my fault. In the Tapajós, I had so wanted to touch the botos. But here, we had not intended to catch the dolphins, only to briefly confine them; yet this near tragedy, which could have drowned two botos, was, I was sure, a consequence of my own desire, flaming like a fire on the edge of the horizon. Again, I am reminded of the damage an outsider can wreak.

My sneakers are encased in mud. My feet are made of clay.

On the way back, we stopped at Curucu. Even after only one day away, we missed "our" dolphins, and although Keila and Socorro knew we were swimming with them daily, they had never seen such a thing and were eager to watch.

Within three minutes of our arrival, the dolphins appeared. It felt like coming home. I swam out to them, perhaps a quarter mile. All seven botos appeared, blowing, pulling their heads from the water to look. A single tu-cuxi leapt. Valentino surged out of the water, showing his flippers. One of the medium grays, Gary or David, rolled on his back, waving his flippers, and then turned, flipping his tail.

Meanwhile, Gilberto and Boanares set the net beside the boat. Within an hour, it was heavy with fish. Dianne, Socorro, and Keila stood on the roof of the *Boanares*, watching—and suddenly, fast as an arrow—"*Rápido! Rápido!*" shouts Socorro—Valentino shot by the net upside down and plucked a fish from the mesh. Another dolphin—which I couldn't see—tried to grab it from his lips. Valentino rocketed to the surface, shaking his head with the fish in his jaws, and jetted away.

We waited to see if more dolphins would feed from the net, but none came. When we retrieved the net, we noted with horror that it contained two four-inch, red-bellied piranhas.

Periodically, we would take the bus into Santarém to change money. It was usually an enjoyable ride of an hour and a half or so, depending on the route, and we became such familiar passengers that the robust, mustached bus driver began to greet us (particularly Dianne) with a hug. Brazilians seem to try to fill every silence with music, and, like most buses, this one came equipped with a large tape deck. Even though we would have preferred Caribe or boi-bumbá, our driver often played American artists in our honor. On this December day, as we return to Alter do Chão from some morning errands, he plays an Elton John Christmas tape. Outside, it is snowing: huge black snowflakes swirl through the windows, the cinders of a fire whose heat we can feel nearby.

We are driving through thick smoke. Our driver is unperturbed. On one side of the bus, the trees are burning. None of the passengers seem concerned. An entire side of the road is flaming like a Yule log, and absolutely no one cares—until we come to a corner where one of the fires chews through the thatched roof of a house. Now they gasp and cry, "*A casa! A casa!*" and crane their necks to see. . . .

As we walked down the sloping red dirt street back into town, we met Keila. We had planned to visit Curucu one last time that afternoon and to take Keila with us. She told us to meet her at two, by the boat belonging to her friend Simão, the *Sousa*. She had a surprise for us.

We waited by the *Sousa*, a larger boat than Braulio's *Gigante*. Then, coming over the gentle slope of the sand dunes, we saw Keila, her blonding, curly hair flowing beneath a wide-brimmed straw hat—and behind her, eight of her friends, carrying bags and baskets. They had come, she explained, to perform for us the Dance of the Boto.

They were beautiful, young, lithe women and men: Edson, Nilson, Carisson, Trenato, and Edany, Elailene, Franides, Diolene, the boys wearing baseball caps and muscle shirts, the girls in little black bikinis and pareos. In their bags, they carried their costumes and unlit torches and props. They had brought a cooler full of snacks and drinks.

The *Sousa*, gap-toothed like its owner, spewed smoke and fumes and noise as we chugged to Curucu; Simão bailed water with the dried husk of a gourd. The boys sat in front of the boat with Simão and Keila, while the girls giggled in the back.

Keila's arm shot out as we approached the point. "Boto!" It was Valentino. As the dancers debarked, I swam out to him, and soon I saw they were all there, as if to bid us farewell: David and Gary, Mary and Jesus, Valentino and Pink Lips, Vera and two tucuxis. In the first thirty-five minutes, I counted twenty-six sightings. David and Gary swam very near, pulling their heads from the water to look into my face. Twice I was enveloped in clouds of sizzling bubbles they sent up to embrace my skin. Three of them leapt in a starburst. David leapt high in the air, not ten feet from me; a tucuxi flipped his tail. Valentino came near and rolled over, showing his pink belly.

I did not realize that on the sandy banks of the river Keila and the dancers, Simão and his little daughter, were all watching in astonishment. When I came out of the water, the strongest-looking boy, with two earrings in one ear, asked me in amazement, *"Voce não tem medo?"* ("You are not afraid?").

"Não—eu gosto!" I replied. ("No, I like it!")

Twice more I entered the water and swam with the botos. I recorded forty-nine sightings in forty-five minutes—and Dianne, on shore, saw many more surfacings, for as usual, without my knowing it, they had surrounded me. And then, it seemed, they moved off.

It was sunset, and a huge fire smoked in the western sky. The dancers were waiting for us on the other side of the point. Keila had set up a little grove of driftwood as props. "My mother doesn't want us to cut trees from the forest," she explained. The young dancers had consumed their drinks and snacks and carefully stowed their garbage away in the cooler. Already, they had donned their costumes. The girls, serious as virgins, wore simple white gowns. They stood facing the river.

The boys, wearing white pants and shirts and hats, began a confident, sinuous saunter toward them, as if swimming. In unison, they began to sing:

> *Quando o boto virou gente*
> *para dançar no puscirum*
> *trouxe arco, trouxe flecha*
> *e até muiraquitã, e*
> *dançou a noite inteira com*
> *a bela cunhantã.*

> When the boto transformed to a man
> to dance at the village ball,
> bringing the lucky bow and arrow,
> and carrying the frog-charm for luck and passion,
> he danced all night with the beautiful peasant maiden.

They danced on the beach like lovers, each couple a unit, tenderly cupping the face or touching the hair of the partner, staring into each other's eyes. "They were blind for one another," Necca had said; in this moment, each partner seems to glow with passion. With the sun at their backs we could see the shadows of the girls' little black bikinis beneath their white dresses, all innocence and new desire. The boys laid down their partners in the sand and embraced them, singing:

> *Um grande mistério na*
> *roça se faz,*
> *fugiu cuhanta*
> *com um belo rapaz . . .*

> The mystery throughout the land
> being the disappearance of the village beauty
> with the handsome man . . .

The sun was melting a golden river on the water, backlighting the dancers. Their features were now indistinguishable. No longer were they these boys and girls; by the magic of their story, they were transformed. Now their song seemed to merge with the voices of the spirits: the Mother of the Vine, who gives power to the plants shamans use on vision quests; the Mother of the Lake, who protects the fishes and dolphins here; the angels of Music and Opera depicted on Manaus's fabulous opera house, the ghosts of divas and murdered Indians. They called out to us to remember. They sang with the voices of the visionaries: those who create and sustain reserves for both people and wildlife, like Tamshiyacu-Tahuayo and Mamirauá, urging us, even in this era of need and of greed, to hope. They sang with the voices of Violetta and Alfredo, the impossible lovers of *La Traviata*. And they sang in the voices of the river guides and the fishermen we had met, Moises and Ricardo, Don Jorge and João Pena—reminding us that, here in the Amazon, the most preposterous of impossibilities can come true.

In this way, Keila's young friends became the story, as ancient, as perfect, as distant, and as present as the sun on the water. And perhaps the dolphins knew this, too. When we looked toward the water, we saw the botos had returned. They swam low and slowly, blowing softly, their eyes above the water, watching the drama on the beach.

> *E o boto ligeiro na roça fugiu.*
> *Desejando a cabocla*
> *na beira do rio . . .*

> The boto fled through the fields,
> desiring the maiden
> by the river's edge . . .

The boys sprang to their feet and, with amazing speed and suppleness, leapt into the river and porpoised through the waves—exactly like dolphins.

We are far from the Manaus opera house and the Meeting of the Waters. Now the air is thick with smoke instead of rain. Yet the journey ends as it had begun beneath the fabulous ceiling frescoes of the Teatro Amazonas: again, I am flanked by Dance and Tragedy.

Out of the water, the dolphin-men emerge. Joyously, each joins his lover, reenacting the promises by which we know the fullness of the world. The botos swim, the dancers dance. But in the western sky, the Amazon is burning.

Lessons Learned
from a Dolphin in Brazil

In March 1994, a friend of mine from São Sebastião, São Paulo's north coast in Brazil, sent me a letter. He was the photographer for the local newspaper. The letter contained a black-and-white picture of a breaching bottlenose dolphin. He wrote: "Dear Marcos . . . enclosed is a picture of a dolphin near the ferryboat pier. Have fun!" This marked the beginning of a unique story.

Two weeks later, I went to São Sebastião. As a dolphin biologist, I knew it was unusual to see bottlenose dolphins close to the shore in this area. So there was a mystery in the air . . . and then in the water as well! It was a juvenile male bottlenose dolphin with an easily recognizable dorsal fin. After seeing how friendly the dolphin was around swimmers and boaters, we decided to write an article about him in which we would ask people to respect the animal and refrain from feeding and petting him.

Two months later the dolphin was still there, and locals gave him the name "Tião." The dolphin was clearly eager for closer encounters with people, frequently approaching them and even touching them. Tião was certainly not the only lone, sociable dolphin in the world, but, as we would see, his encounters with people would be unusual.

In August, Tião moved northward to the city of Caraguatatuba, several miles from São Sebastião. His encounters with swimmers were more frequent. Swimmers tried to grab on to Tião's dorsal fin for a ride, imitating behavior they had seen on the television series "Flipper." The reality, however, was that Tião was a wild animal unsuited to such tricks. The dolphin naturally tried to free himself when this occurred, and the local hospital received some swimmers with cuts and abrasions as a result. The situation was critical. The austral summer came in November with all the beaches crowded and the hospitals busy with injured swimmers who had tried to play with Tião despite our warnings.

In early December, Paulo Cirilo, a close friend from Caraguatatuba who had been monitoring this situation, called me, saying he was desperate for help. I told him I would seek assistance from the federal environmental agency's main office in São Paulo. I had no success. Their reply was disappointing: "Swimmers must know that the dolphin is in its element and that humans are not always welcome there. They must learn who is strongest in the water." Paulo insisted that somebody could die if specific measures were not taken. He told me he would try to put some cords and buoys in the water along local beaches that might discourage the dolphin from approaching the beaches and the swimmers.

Paulo was right. Tião avoided the beaches everywhere except at the Martim de Sá, the only beach that was still without buoys. On December 8, 1994, that beach was crowded with tourists. They wanted to swim with the famous dolphin. Some of them were drunk. Tião was there, surrounded by twenty-five to thirty swimmers. Some of them abused the animal, trying to jump onto his back or tie him up with strings, even to put ice cream sticks into his blowhole. After several hours of such interactions, a few swimmers had the idea to take the dolphin out of the water to photograph him. Three of them managed to drag Tião out of the water. While doing that, the dolphin flailed wildly to free himself and hit a swimmer's belly in the process. While others continued to play with the dolphin, the injured man lay motionless on the sand. Concerned friends took him to the hospital, where they were given grave news: the man had died from internal bleeding caused by a stomach rupture. It was apparently the only known account of a human fatality due to an encounter with a free-ranging dolphin.

In the days following this tragic event, the newspapers published shocking headlines: "A man was murdered by a dolphin"; "Killer dolphin on our beaches"; "Dolphins are not graceful anymore: They are killing us." A nightmare had begun, and the news spread all over the world. Some people wanted Tião dead. Others said that the accident would teach humans to respect wild animals. Some people wanted to capture and train the dolphin for public exhibitions. Others wanted to relocate Tião far from the shore. Killing him to expose his skeleton in a local museum was one of the worst ideas. This was the chaotic situation we lived.

We had never faced a problem like this in Brazil—in fact, no one ever had. Perhaps because Tião had no other companions, he remained accessible to swimmers even after this incident. Because there were no Brazilian laws to protect Tião, we feverishly went to work in three different ways: through public education, media control, and with the prevention of harmful interactions between swimmers and Tião. Twenty thousand brochures and fifty advertisements were produced. A crew of nine members was trained to educate swimmers on local beaches. On a weekly basis we gave press releases to the media. From a distance, we would follow Tião and ask swimmers to go to shallow waters to give him space when he approached.

On five occasions in the summer of 1995, far from swimmers' attention, I got into the water with Tião. I thought that by "meeting" him I might gain a helpful perspective on some of the accidents that had occurred. When I was in shallow waters, with my feet on the sandy bottom, Tião tried to gently knock me off my feet with his body. If I did not allow this, Tião then used his open mouth as a tool to gently pull at my knees. After removing my feet from the floor, Tião pulled me to deeper waters. Here, he vocalized much more than in shallow waters, swimming very close to me, diving with me, and even breaching on some occasions. After those interactions I understood more about Tião's life in his underwater world. Tião remained in São Sebastião waters until August 1995, when he was last seen. We still do not know what became of Tião—whether he simply avoided people or had died. I will always wonder what became of him.

Tião's story is a dramatic example of the complicated and unpredictable nature of the human-dolphin bond. Although we may never know how this story ends, its unfolding has taught us much about dolphins and humans and the boundaries that lie between us. I take comfort in the fact that each new dolphin encounter provides us with a chance to learn more and that, perhaps, one day, we may reach a balance between the species.

Luminary

L ike the whales and dolphins who ascend to the surface, retrieve a breath, and descend once again through the fathoms of the oceans, I have been carried to wondrous heights of inspiration and great depths of experience by these warm-blooded beings of the water world. They have permeated my thoughts, artistic expressions, scientific pursuits, and dreams. They are my reason for having spent many hours cradled aboard seagoing vessels and standing upon the shore, a researcher, an ally, a guardian looking out for their safety. For more than a decade my work has taken various forms, but always it has embodied the same ideals. As whales and dolphins become increasingly threatened by the human world, I have endeavored to safeguard their physical bodies. I have also sought to secure some haven for that part of them which is no more tangible than my own emotional and spiritual self. I aspire, as have so many people through the ages, to grasp some quantifiable or perhaps acutely intuitive insights into who these creatures of the sea really are.

In the presence of dolphins and whales, I have experienced moments of clarity that have revealed much more than the tremendous triumph of natural selection and the biological genius these creatures exemplify. And though I can offer no proof, it is during these precious moments when my consciousness has awakened to the mingling of our souls that I have known an astounding grace, humility, and cause for jubilant celebration. These are powerful moments, which can no more be captured than the surging, changeable oceans of which the whales are a part. These are the moments that continue to motivate me to pursue the protection and discovery of these beings who, in their naked innocence, cause us humans to bare our own souls. We, like they, are both vulnerable and powerful, the sum of much more than what can physically be observed. And the degree to which we honor this unseen essence may determine our own fate and theirs. It is with these thoughts about the essence—the spirit—of whales that I invite you to share in an excerpt from my journal.

On a chilly Newfoundland evening, I set out to document whatever I could of the acoustic expressions of the second of four solitary sociable beluga whales I have been privileged to meet, study, and protect so far.[1] I remember an air of curious anticipation. Perhaps it was the ethereal nature of the serene mist gently settling into shadowy treetops around the small harbor. Or maybe it was the fact that everyone else in the village of King's Point had closed their doors on the darkness, and out under the vast open sky I was alone, except for the whale and whatever other creatures might be casting about in the nearby forests and sea.

I was confident that my video camera would do all the official recording of acoustical data for later analysis. So, I decided to depart from the usual routine of documenting on data sheets and, instead, write freely in my journal. I knew there would come a time when the little whale would move on and the likelihood of ever seeing him again was not good. In the future, I would want to recall every precious moment. And so, amid time/event reference notes, I wrote virtually every rambling thought, memory, and feeling that surfaced. I hoped that, buried within my scribbles and unrestrained lyrical musings, I might find some clue to the insights I constantly seek.

September 16, 1999
Kuus—100 yards from the wharf

It is coming up to 21:00 hours and this will be the final observation session of the day. I am eager to hear what goes on at nighttime in the little whale's world. I am huddling on the edge of the weathered boatyard jetty, here at the head of King's Point harbor. My camera is balanced atop one of the pilings. I just plugged the hydrophone into my Sony Hi8, and as I slid the sensitive microphone receiver into the water it registered a soft "plunk." Everything is working. With my headphones pressed snuggly to my ears, I can listen to the resonant world below the rising tide. I'm ready to begin scratching time code notes and referencing every nuance and sound that I hear the little whale make.

20:55:25   *creaky door—distant, echolocation*
Ah, there! A slow, crisp staccato—kga . . . kga . . . kga . . . kga . . . kga . . . —
starting in a low baritone and rising in pitch as each click pops ever more rapidly
and runs one into the next, until, in my ears, they form a continuous, whining buzz.

20:55:40   *whistle!*
Much like that from a bird in a tropical forest, the notes soar skyward on a mu-
sical scale that threatens to breach the threshold of my limited range of hearing.
Beluga sounds are so wonderful! Our verbal descriptions of their vocalizations just
don't do justice.
The pages of my journal are already getting damp and warped by the cool night
air. It's mid-September and, as summer wanes, the last of the season's tenacious
mosquitoes are hovering about, greedily taking chunks from my ankles.
A single flood lamp dimly illuminates the boatyard wharf. A faint splash of light
spills over the edge onto the calm dark water below, allowing me to glimpse schools
of small fish passing by at the surface. Capelin, I think. I've been told it is rather
late in the year for capelin, but then there have been so many unusual occurrences
in these waters of late. The beluga whale is just one of them. I wonder if Kuus has
gone after these small schooling fish or if he is in pursuit of larger prey?

20:56:49   *creak—distant*
Above the surface all is quiet. The beluga whale is out of my limited sight, but
the cold, languid ocean is alive in my ears. It's thrilling that even in darkness I can
peek intimately into this other world. I feel like a kind of auditory voyeur listening
to every sound as Kuus comes and goes . . . in and out of my hydrophone range.

21:01:56   *click, click . . . echolocation—slightly closer*
It sounds like the little beluga is up the harbor about halfway to the village.

21:02:37   *creaky door . . . echo*

21:03:19   *squeak, echo—fainter*
It's so amazing that I am here, studying yet another solitary sociable beluga whale!
Who would have thought that after "Wilma" there would be another one, and only
a few months after she disappeared from her adopted waters in Nova Scotia. I will
always be indebted to Wilma. She was the reason I founded the Whale Steward-
ship Project to research and protect whales and educate the public. Now with Kuus,
a two-to-three-year-old beluga orphan, it's like being given the chance to go back
in time to when Wilma first arrived in Chedabucto Bay six years ago. Field re-

search isn't often so generous. Kuus probably arrived from the Arctic and is about the same distance away from home as Wilma was, hundreds of miles from her family group in the St. Lawrence.

   21:05:10   . . . ?—*whale sound or maybe a squeak from a mooring across the harbor*

With all her irresistible beluga charm and growing trust, Wilma fell prey to human activities that were frequently misguided, clumsy, and irresponsible . . . Am I being too kind? I think that most people certainly meant her no harm, but everyone wanted "a piece" of her. In the process, fragments of her body were literally sliced away by propeller blades whirring mercilessly, even if by accident, into her flesh. It was her scars, sketched from head to flukes, that people always pointed to in acknowledgment of the profound need for the Whale Stewardship Project.

I really don't want history to repeat itself with Kuus . . . and so far, so good. He did have one collision with a boat propeller before I got here, but now something quite delightful seems to be happening. People are actually choosing more often to come out in their rowboats, using their own muscle power and paddles instead of horsepower and menacing blades. They naturally want to get close to this cute little fellow but are really being careful not to hurt him. I know it can be a challenge sometimes, putting the whale's needs ahead of our own comfort or convenience. I've endured more than one vigorous workout battling northeast winds in a rowboat or canoe. Thankfully we are a long way from the open sea, or else they might still be searching for me out there.

   21:08:25   *creak, echo—faint*
I wonder how far from the hydrophone Kuus actually is. . . .

It does make me feel very good, the way people here are so diligently applying the Whale Stewardship Project guidelines for interaction with solitary sociable beluga whales. I only wish I could have been there sooner for Wilma. By the time I had initiated the protection and education program for Wilma, she had already been resident for five years—a very long time for treacherous habits to become ingrained in both the whale and humans. During the recreational season, the numbers of people who wanted not only to see her but touch her, too, had grown into the thousands. A multitude of hands reached out, poked and prodded the trusting little whale. And when Wilma did not "come" like some kind of obedient domesticated pet, people impatiently slapped the water or rapped insistently on boat hulls. Much worse, some even brazenly exploited Wilma's chronic attraction to the sound of boat motors, ruthlessly switching them on and off to entice and manipulate her. Oh, what a summer!

I had no legal enforcement authority, but I did scrape together just enough public and corporate funding, a research permit from the Department of Fisheries and Oceans, and their blessings to offer my unofficial guidelines. Without a doubt, it was about the most extraordinary and rewarding adventure of my life. And it hasn't stopped. . . .

Where is Kuus? I haven't heard him for several minutes.

To my knowledge, Kuus has really only been living this close to humans for a few weeks, so initiating our program promptly is, I'm sure, the greatest reason we are meeting with such success. Since Kuus is clearly attracted to engine sounds, just like Wilma was, our guidelines state that people should switch their boat engines off when they approach the vicinity of the whale. So far, most everyone is respecting that. Thankfully there aren't as many people trying to swim with Kuus, and no one has attempted to ride on him either. Swimmers often tried to grab on to Wilma as if to realize some fantasy where they could emulate marine park trainers performing a circus stunt—hardly a healthy, safe, or respectful conservation ethic. These little beluga whales are all alone, not something that is normal for them, especially ones who are barely more than babies. Their need for meaningful companionship is innate, compelling, and makes them terribly vulnerable. Put that together with human genius for exploiting a situation, and sooner or later the "resource" is severely endangered. Wilma certainly paid the price for companionship. She endured hands, feet, and even someone's head being stuck into her inquisitive, gaping mouth in addition to all her injuries.

We can observe bodily scars and acknowledge physical pain. But what other kinds of wounds had Wilma suffered—what psychological and emotional torments from the absence of other belugas in her life and from the overwhelming presence and demands of humans?

21:17:02   *echo*

21:17:32   *echo—still faint*

Most assuredly the daily presence of our Stewardship boat in Wilma's habitat did make a tremendous difference, and, to be fair, there were plenty of people who went out of their way to show the utmost respect for her. I was always heartened by those momentary interludes when physical contact between Wilma and humans, especially children, was ever so gentle; a tender exchange with an underlying connection that belied the vast differences between our species. Yet, I feared even these special moments might in some way eventually be harmful to the whale.

I think Kuus is really benefiting from people generally giving him more space and, perhaps most of all, allowing him the option to choose and control his interactions with them. I feel certain it is because of this special respectful approach that I have already observed some amazing interplay. Yesterday, Kuus brought presents of seaweed up to David and Linda, two of his consistent and caring friends. While carefully balancing in their canoe, they received the whale's gift and handed the kelp fronds back and forth. These encounters unveil wonderful glimmers of the potential for our relationship with whales. If only humans could resist the temptation to exploit it. More than anything, the exchange reminded me of my yearning to know intimately the enigmatic part of whales that is their essence.

21:20:25   creak, echo

I have only known this little whale for a few weeks, but with each day out here studying and working to give him every advantage, Kuus becomes more and more a part of my life. Each hour that I spend observing his every movement, listening to his every utterance, I feel more deeply drawn into his life—so unique, so individual. What an extraordinary gift to have this whale become inextricably woven into the fabric of my own being.

I chuckle when people ask "Do you think the whale 'knows' you?" I suppose, from their perspective, I do spend countless hours, sometimes in the driving wind, rain, snow, and freezing cold, out here "with" the whale. Oh, I am certain that he has the ability to recognize and differentiate between people, given enough time, exposure, and relationship-building. But for me it is quite different. This is very much a one-sided love affair. I explain that our research protocols require that I remain a secret admirer: observing every detail of his life while remaining as separate and unobtrusive as possible. In fact, I actually must become so boring to Kuus that he simply carries on about his business even though I might be seated in a boat or on a wharf less than a few feet from him. Hmm. I admit that there are times when I long to reach out to caress his buttery-soft skin. There are times when I just want to be crazy and fun! I want to celebrate this amazing little guy. I want to sing and squeal and make strange and wild noises, just to see if he finds anything I do interesting. And I want to reach out in other ways too. Sometimes it breaks my heart to ignore his invitations. His overtures to "play" or otherwise interact can be so clearly communicated that when I do not respond, well, I'm sure he must think that at least this human is incredibly stupid! But I do understand that he wants companionship. And I do understand that his need to relate and express is a huge and integral part of who he is as a beluga whale. But I also understand that there is a bigger picture, where he needs, more than anything, to reunite with other beluga whales, and we humans would be doing few favors for this little fellow by attempting to provide surrogate companionship. Though so often my heart aches just to be his friend, for his sake I am resolved to set an example for other people concerning the most responsible and respectful way to behave in the presence of this whale.

21:23:40   *echo*

21:24:00   *interrupt data collection for battery change*

21:26:34   *echo—closer*
The headphones are keeping my ears warm but my nose and fingers are starting to feel the chill. The sounds below the water have been surprisingly few and rather faint. Most prominent is the periodic crackle and snap of crustaceans and the grunting sounds of some kind of fish I have yet to identify. A few larger fish are flopping at the surface.

21:27:57   *creak, echo—close creak, high pitch chirp*

21:28:06   *echo—very close, surfacing maybe thirty yards from wharf*
Oh, this is great! Kuus is much closer now, and the hydrophone is really picking up his vocalizations and echolocation.

21:29:17   *groan . . . echo . . . wheeze*
The sound is so clear, it's as if he is right inside my head! I hear the clicking and tap tapping of his echoes. Buzzzzzzz . . . This little whale is deftly finding fish and honing in on his meal . . . such a welcome sound. He's making a living.

I listen intently, my mind bursting in its attempts to translate the reverberating click-trains into mental pictures that I can comprehend. I imagine his invisible emanations of sound in the deep dark water, leading his grinning mouth ever closer to the prize—a sumptuous fish. I am humbled at the realization that my own senses are so limited. With each pulse of his echolocation I am reminded how infinitely superior this whale is in his environment. I feel a twinge of envy. Or is it a longing to have the mystery revealed, if only for a moment, of what it must be like to transmit sound from my head and have it return to me like a boomerang full of information and images of what surrounds me? I am in awe of this being who can "see" with unrivaled clarity and precision into the blackness of the night's watery depths.

21:36:56   *whistles and squeal!!! squeal!! squeal! squeal!*
Oh, my dear God! This haunting cry, screeching in my headphones—it is breathtaking!

21:38:15   echo and squeal!!! whistle
I can feel my heart pounding. I truly was not prepared for these calls to come exploding through the periods of silence and rhythmic echoes. If only there were other beluga whales out there somewhere within earshot! How many a night has this little lost whale called out through the inky waters? And nothing. No familiar chirps, whistles, or trills to reassure him. No big, warm, pudgy bodies just a pectoral flipper away, to rub up against for comfort and security.

21:38:29   whistle, squeal, echo!!!
Little whale . . . mortal, thinking, feeling sentient being.

21:38:48   squeal!!! echo
What message is contained in his plaintive cries that vibrate mercilessly within the center of my brain, piercing my soul and every fiber of my being? Until now, I had only guessed at the pervasiveness of his aloneness. Here, on the wharf at the head of the bay, the darkness seems more bleak than ever as my senses take in the heartrending howls emanating from the deep dark water.

21:40:27   creak

21:40:52   echo

21:41:13   creak
There is no one here to hear him. No one except me. The heaviness in my chest is erupting, welling up. I cannot—do not—hold back the tears that spill so effortlessly from my lids and cascade down my face. I cannot reach him. Cannot tell him that he is not entirely alone. I, in this other world of air and earth just above the surface, am listening. "I hear you Kuus. I want you to know that I hear you. . . ." My own feeble muffled voice trails off as, silently, my mind finishes ". . . somehow." The pragmatic side of my brain mockingly points out the futility of the thought.

I am a poor substitute, I know. Still, I can't help feeling that he might find a tiny bit of comfort in knowing that some other being, strange as I may be to him, recognizes him and is reaching out. Attempting to cross the boundaries that separate his species from mine. Boundaries far more impregnable than those that currently separate him from his family pod.

21:50:29   groan, creak, echo—close!

21:51:09   echo—close, whistle, squeal!!! creak, echo

21:52:35   surface and blow, echo

All is quiet . . .

Breathe, breathe deeply . . . like the whale.

Little fishes brush indelicately against the hydrophone cord, sending crunches and crackles into my earphones. Stars are trying to break through the faint mist that shrouds the crescent moon hanging in the west. Ah, a Kuus moon.

22:11:45   call (squeal)—distant

22:17:22   echo, creak

22:18:42   echo—distant

22:21:04   squeals—distant

I hear Kuus off in the distance again. Kuus's friend David, who often ventures out for a visit in his canoe, did a little research and suggested the name I have adopted for the beluga. "Kuus" is a Beothuk Indian word, meaning "moon" and "luminary." Their symbol for the crescent moon, like the one in tonight's sky in fact, was sacred to the Beothuks. It is a fitting image for the beluga whale whose finless back curves gracefully at the surface as he rises while traveling and filling his lungs with air. Yet, I think even more profound is that this little whale, who is enlightening so many people in the towns and villages of a remote part of Newfoundland, is called "luminary."

22:30:10   echo—distant

22:32:08   call (squeal)—very distant

The dampness is invading my bones. It has been several minutes since I last heard Kuus's faint call in the distance. I've been listening in the dark for almost two hours.

A light breeze is beginning to wrinkle the mirror-calm bay. My camera batteries are nearly drained. I'll soon be leaving the wharf. My toes and fingers will be glad of a little warmth. The thought of a hot steaming cup of herb tea followed by the flannel sheets and cozy comforter on my bed sound inviting. But, I never know when Kuus, like Wilma, will simply disappear. I cannot go yet. I am determined to stay and eavesdrop on the underwater world of this beauty with the delphinic smile and deep beckoning eyes, until the last drop of juice is squeezed from my battery packs.

22:42:10   echo

22:42:58   echo

Kuus is back. My batteries are emptying quickly. Darn!

22:43:41   *echo, surface and blow, echo*

22:43:59   *echo and creak—close!*

22:44:50   *creak + echo + creak + creak + echo + creak!*

22:45:50   *blow, echo, groan . . .*

Silence! It is deafening. I feel cut off. Disconnected. I feel disembodied from my kindred whale spirit, no longer able to hear his calls and echoes against the backdrop of the other sounds in his liquid world. Without the aid of my camera, headphones, and hydrophone I am deaf . . . acoustically separated from this incredible teacher, mentor, and, dare I say, friend.

Now I can hear only his blows when he comes to the surface. Only his breaths so infused with life and determination punctuating the night air and assuring me he is still here, traversing his world, making sound pictures I can only imagine. He dives and all is quiet again. Cold as I am, I can't seem to shake my reluctance to leave. I hover here on the wharf, shivering as I try to assimilate the thoughts and feelings, the awe and wonder, the powerful sorrow and the sublime exhilaration I have experienced through this transportive merging of my soul with the essence of this whale.

It is eight minutes to eleven. I am going in to get warm.

As mysteriously as he arrived in the waters bordering some of Newfoundland's most beautiful coastlines, Kuus vanished—but, like Wilma, not without leaving in his wake many powerful memories, lots of behavioral data to study and analyze, and not without inspiring my heart and imagination.

There are few things more thrilling than being an explorer and preserver of the body, mind, and spirit of a solitary sociable beluga whale. It seems, with every moment in their presence, I edge closer to knowing who they really are. And after all the logical, quantifiable, and statistical analyses are done, I only hope that I may adequately convey that invisible, invincible spirit these whales are so generously showing my heart.

NOTE

1. T. G. Frohoff, C. Kinsman, N. A. Rose, and K. Sheppard, "Preliminary Study of the Behavior and Management of Solitary, Sociable White Whales *(Delphinapterus leucas)* in Eastern Canada," paper presented at the International Whaling Commission

Conference, Subcommittee SC/52/WW3, in Adelaide, Australia, in 2000; C. Kinsman, T. G. Frohoff, N. A. Rose, and K. Sheppard, "Behavior and Occurrence of Solitary Beluga Whales *(Delphinapterus leucas)* in Eastern Canada," in *Abstracts from the 14th Biennial Conference on the Biology of Marine Mammals* (Vancouver, B.C., Nov. 28–Dec. 3, 2001).

CHRISTINA LOCKYER
AND MONICA MÜLLER

Solitary, Yet Sociable

Why do we find solitary dolphins that are sociable with humans? The solitary state is often temporary, perhaps triggered by the dolphin's loss of a companion or group. Maybe prey availability has changed or habitat destruction necessitates a move to a new area. It is probably incorrect to assume that such solitary animals are misfits that cannot function in dolphin society; at some point these dolphins may rejoin their own society. The sociability evolves because of solitary circumstances and prolonged exposure to human activity. The dolphin's innate inquisitiveness will encourage it to explore interactive possibilities with other creatures.

Most cases of solitary and sociable dolphins have been the bottlenose species,[1] and the sociability factor may be influenced greatly by the fact that the bottlenose dolphin is generally an inshore coastal species where contact with humans and their activities is unavoidable. To date, about seventy solitary and sociable dolphins have been recorded worldwide—the first proper documentation perhaps being "Opo," a young female off New Zealand in 1955–56.

Many people familiar with solitary and sociable dolphins think that such dolphins are very different from those dolphins that live in groups and do not approach people. This common belief about sociable dolphins is not surprising if we look at examples of the behavior of some very sociable individuals.

THE QUESTIONS

Do all so-called "solitary and sociable dolphins" allow similarly close body contact and actively seek human companionship? Are they, from the moment of their appearance at some marine coast, bay, or harbor, just different from their "wilder" conspecifics who do not seek human contact? And do sociable dolphins really depend on having regular interactions with human swimmers? These questions are rarely asked, and few precise answers are known today. But if we try to compare the behavior with humans in different known sociable dolphins, we may find interesting insights concerning these questions and, as a result, reach a better and deeper understanding of the very special situation in which these animals live during their time as "solitary and sociable dolphins."

Here we share with you the differing stories of solitary and sociable dolphins we have known personally off the coasts of Europe. In France, there were five females who became familiar with people and well known by the media.[2] Only one, "Françoise," remained in the same area until recently but was no longer sociable with humans. Regrettably, Françoise died in the summer of 2001, entangled in fishing nets that have been used in the Arcachon lagoon. There is no solitary and sociable dolphin left in this lagoon now. Of the sightings in the British Isles, individuals have appeared off the eastern Scottish coast, the western coast of northern England and Wales, southwest England, and Ireland. We will present three individuals, all males.

Jean Louis

"Jean-Louis" was first reported by a local fisherman off Cap Finestère in Brittany in the spring of 1976. The local people mistakenly believed she was a male, and so Jean-Louis became famous with her male French name during her twelve-year stay in the same marine area.

The dolphin used to follow local fishing boats in this area, in particular to inspect fishing nets and lobster cages. Initially, between 1976 and 1982, Jean-Louis became increasingly famous in Brittany and regularly had boats and swimmers trying to approach her. At this time, she often remained several meters away from swimmers or observed them from some distance while accompanying them during swims or dives.

Only a few people could approach this dolphin closely, but she did not allow bodily contact. In 1983, huge numbers of people from all over France and from abroad visited Jean-Louis's bay to see the famous dolphin. However, Jean-Louis seemed to react to these crowds by showing much less interest in swimmers than before, yet she still regularly followed her "favorite boats" around. Until 1988, she had some very intense interactions with special swimmers—yet would completely avoid others. She also touched some swimmers by rubbing her body against their legs or bodies. The dolphin always initiated special "games" with swimmers and decided the course of interactions. In December 1988, Jean-Louis disappeared suddenly. Nobody knew where she went or why she disappeared, and, so far as we know, she was never seen again.

Fanny and Marine

The story of the subadult female dolphin "Fanny" and her human protectors began in the spring of 1987 when the solitary animal took up residence in a very restricted marine area at the Cap Couronne (entrance of the Gulf of Fos), east of Marseille. Apparently fishermen first observed her circling continuously and exclusively around an immense buoy, when she was only 2.5 meters long and estimated to be about five or six years old. During this period, Fanny showed some interest in boats but remained very distant from swimmers. Nobody could approach her closer than four or five meters.

On September 15, 1988, Fanny was joined by "Marine," another female bottlenose dolphin. Both dolphins rapidly became inseparable and mostly swam in unison, blowing and diving synchronously. From the very big belly of Marine and the behavior of the two dolphins, it seemed obvious that Marine was pregnant. Members of the "Observatoire Fanny" believed that she had been looking for another female dolphin to assist her with pregnancy and birth. Fanny adopted a sort of protective behavior by always placing herself between Marine and swimmers, divers, and boats. Nobody could ever approach Marine closely during this period. During the winter of 1988–89 the two dolphins disappeared for several weeks after a heavy storm and were not observed again until March 1989.

When they returned to the area, it was clear that both dolphins had become victims of serious harassment: Fanny had been wounded, and Marine had aborted as indicated by her flat belly. After this bad experience, both dolphins appeared very anxious and much more distant than before, often totally avoiding contact with boats and swimmers. In May 1989, Marine left Fanny and disappeared from the Marseille area.

In September 1990, Fanny appeared for the first time in the very pol-luted harbor and channel of Port-Saint-Louis du Rhône, and she did not leave this artificial area until her final disappearance in 1994. Among the people drawn to Fanny in her new environment was one particular family with a twelve-year-old daughter, Sylvia Chambon, who succeeded in gain-ing the confidence of the dolphin by swimming next to her every day, even during the worst weather. Slowly the dolphin accepted body contact with the young girl, offered her dorsal fin to draw Sylvia through the water, and dived with her in unison. During four years in the area, no one else succeeded in approaching Fanny so closely. When Fanny finally left this area on May 8, 1994, she was an adult female of about three meters. She was never seen again, like so many other solitary and sociable dolphins after they disappear.

Dolphy

When observed alone in the spring of 1990 by the employees of the aqua-culture farm in Baie de Paulilles, "Dolphy" stayed many months without allowing close interactions with divers and swimmers. During late 1990 and 1991, she began to allow very close approaches by divers and swimmers and also began to swim regularly with a little dog called "Rocky" who met Dol-phy on the open sea and in the harbor of Banyuls-sur-Mer.

Rocky was owned by Jean-Claude Roca, a professional diver at the Ma-rine Institute "Laboratoire Arago" at Banyuls who went diving nearly every day in the institute boat accompanied by his dog. He met the young dol-phin in different marine areas close to Banyuls. Every time Rocky caught sight of Dolphy's dorsal fin, he jumped excitedly into the water and tried to follow the dolphin, who seemed pleased to play with the dog. Dolphy and Rocky interacted for many hours together, winter and summer, until her disappearance in 1995.

In 1994, we first found teeth marks of other bottlenose dolphins on Dol-phy's skin. That year she also performed her first long distance journey of about 400 kilometers in July by swimming along the Spanish Costa Brava to the huge and very polluted harbor of Barcelona, where she stayed for one and a half months. There again, she had body contact and close encounters with different swimmers and divers and used to follow fishermen as well as excursion and pleasure boats.

In May 1995, Dolphy was observed together with two other bottlenose dolphins who accompanied her every day for two weeks between the harbors of Valencia and Gandia, some sixty kilometers southward. During this time, Dolphy still entered harbors to follow boats and approach swimmers, but the other two bottlenose dolphins remained outside of the harbor entrances until Dolphy rejoined them. She finally disappeared without being seen again.

Françoise

As mentioned above, Françoise was a subadult female bottlenose dolphin of the Arcachon lagoon on the French Atlantic coast (region of Gironde). She developed the characteristic behavior pattern of the "solitary and sociable" dolphin, but only at what could be called a "semi-solitary" level, and so she was somewhat unique.[3] From 1989, when Françoise was first observed in the lagoon, she only followed boats and sometimes approached swimmers. During that year she was still frequently observed together with five other bottlenose dolphins, a group that probably included her mother.

A scientific survey of these six dolphins between 1988 and 1991 by a French marine mammal research group[4] had revealed that this small group constituted the only dolphins who were permanently resident in the Arcachon lagoon. The group remained stable, without any new births, during the ensuing ten years.

At the beginning of the scientific survey, Françoise's behavior when she was away from her group was similar to the behavior of other solitary and sociable dolphins. She always kept a distance of at least one meter from any divers and swimmers. She was also often observed approaching different people in shallow water or swimming and diving around their boats and bow-riding for long periods alongside fast boats and jet-skis. She loved rubbing her body against ropes and playing with buoys and other objects floating in the water.

However, no difference could be seen between the behavior of Françoise and other members of her group when they were monitored in the sandy channels of the lagoon. During these periods of sociability with her group, the young female was generally in the middle of the group, avoiding boats and swimmers, performing very long dives in unison with them, and sharing all other group activities.

In August 1995, "Vire," the oldest female of the group, died. Vire had already been observed alone more frequently for several months before dying. Afterward, Françoise associated much more often with her conspecifics and became much less sociable with boats. Observers believed that the death of Vire changed the status of Françoise within the dolphin group. Françoise thereafter remained mostly in her little group; when alone, she spent hours stationary next to a big buoy, sometimes accompanying boats inside the lagoon but no longer engaging in contact with swimmers or divers. As noted above, she died in 2001.

Beaky

"Beaky" was an old male of 3.6 meters in length. He was very battered around the jaw, with white scars on the tip, and was also easily identified by many other accumulated scars on his body including a healed bullet wound. He first appeared off the Isle of Man in western England in March 1972 and frequented the harbor area of Port St. Mary. He became well known locally and began to approach boats and particular individuals in the water.

In mid-March 1975, Beaky suddenly disappeared, never to return to the area. Coincidentally, this happened after months of harbor construction involving the use of underwater explosives in Port St. Mary. After a few weeks, a dolphin matching Beaky's description turned up in April 1975 in snow and windy weather off the coast of Wales in Pembrokeshire, having followed the local ferryboat *Sharan* from Milford Haven to Skomer Island. The apparent attraction was the small wooden pram-dinghy that the *Sharan* was trailing.

Further investigation confirmed that this was Beaky. He rapidly became sociable and followed boats and swimmers, allowing people to pet him and play with him. He always appeared to recognize certain individuals and would seek them out. Voice, attitude, and general appearance may have been the clues to identity; clothing my not have been so important. After several long excursions away from the area (a few days to weeks), Beaky abandoned the area in mid-January 1976 and turned up several days later farther south in Mousehole in Cornwall, where he stayed close to a mooring buoy off the Penlee lifeboat station.

Beaky's interactions with people eventually included allowing swimmers to take hold of his dorsal fin and giving them a ride. He would also play with inflatable balls and rubber rings that people tossed to him. He liked to approach, rub against, and maneuver boats—to such an extent that he once caused chaos during a small yacht race when he surfaced under different competing boats and turned them around or near-capsized them, to the dismay of the crews! He liked to be petted and would occasionally become sexually aroused. At other times, especially when many people were in the water close to him, he would become aggressive and try to butt people with his snout or leap from the water. He would graciously accept fish offered to him near "his" mooring buoy, but Beaky never ate them. They were played with and then abandoned on the seabed, as witnessed by a heap of rotting fish. Like Dolphy, he enjoyed interacting with dogs that entered the water and would try to chase them in play. What triggered his final departure after August 1978 is unknown. He was never seen with other dolphins, but certainly there were indications that some of his long absences had been related to mackerel migration and prey movements.

Percy

Another male bottlenose dolphin was "Percy." He was a very old and big animal—perhaps one of the largest recorded—at 4.1 meters in length with worn-down teeth. Like Beaky, he frequented an area around the southwestern part of England, but in the northern part of Cornwall between St. Ives and Portreath. He also had distinguishing marks and scars, but none as clear or permanent as Beaky's. Unlike Beaky, Percy frequently bore tooth rake marks—undoubtedly from encounters with other dolphins, but which faded over time. Percy often made excursions away from his home range area for several days, during which time he may have encountered other dolphins. But no one ever saw him with conspecifics.

Percy took more time than Beaky to associate closely with people—years rather than months—but, like Beaky, he was fascinated by boats, especially lobster fishing boats, which he would follow when they pulled and set pots in the area. He would bow ride fast inflatables and even leap clear over them—a somewhat exhilarating experience for the occupants! He was very inquisitive and initially showed interest in commercial divers working underwater by swimming up close and inspecting and mouthing their tools. His first encounter was in fact with a wreck diver in January 1981.

Like Beaky, he also recognized certain individuals and regarded them as companions or "property" that he would protect if others came too close. Unfortunately, this resulted in a few apparent acts of aggression in which Percy butted and pushed a swimmer out to sea, who eventually needed rescue. Percy was clearly a dominating animal who liked to dictate the terms of behavioral conduct. He would play with swimmers but then not allow them to leave the water, and sometimes he pushed divers or snorkelers down against the seabed and tried to prevent them from leaving, as if he had been playing and wanted the session to continue. He would get very excited if many people entered the water at the same time, and he often become aggressive, pushed some away, and tried to stop others from moving away. He even bit people in his frenzy, although this was rare. Also like Beaky, Percy was able to distinguish between men and women; he would frequently investigate the groin area of women. He often singled out menstruating women for his attention and ignored the others, a fact that discouraged women in this condition from entering the water with him. Maybe he could taste traces of body fluids and hormones.

Percy had a great liking for chasing and leaping, and he would follow surfers and windsurfers closely. This also resulted in mishap, with Percy leaping back and forth across surfboards, frightening the people, and once landing on the front of a board and breaking it. The surfer had to be rescued. Another time he harassed a windsurfer to such an extent that the person had to be rescued. Percy's behavior often seemed to be mischievous—an anthropomorphic view, perhaps, but it became clear that people had to learn to give him space and respect the fact that the water was his environment, not theirs. He liked to interfere with mooring lines and was known to lift small anchors and move them around as well as drag lines and boats by pulling on the lines, which he allowed to pass through his teeth like dental floss. Everything in his environment was for investigation and play. He even tossed large sunfish (*Mola mola*) clear of the water and balanced them on his snout.

Although he never allowed people to pet him in the same way as Beaky did, Percy became sufficiently relaxed after two to three years to allow touching, and he allowed us to measure him and feel his scars and teeth. Like Beaky, he liked to rub against boat hulls and people. Unfortunately, during the last few months of 1984, Percy became very interactive and more aggressive toward people. This was likely triggered by the fact that his presence attracted more visitors, so that during these months he was continuously harassed by people and pleasure boats. Perhaps it was just as well that he disappeared in November 1984, before any problems occurred.

Simo

Our last dolphin character is "Simo," a juvenile male bottlenose dolphin. When first measured he was under 2.3 meters but, during a period of only fourteen months, he grew to 2.7 meters. He first turned up in Wales off Pembrokeshire in January 1984, in an area near Solva. Unlike Beaky and Percy, he very quickly adapted to people and their ways after an initial wariness, and he became easily approachable. He was very relaxed with swimmers and divers and enjoyed being petted and stroked. He would also encourage swimmers to hold his dorsal fin or pectoral fins and carry them through the water, often belly to belly, coming back if they lost their grip. He played with objects that people introduced into the water as toys, and he liked to chase boats and come into very shallow water inshore and swim between bathers' legs.

Just like Beaky, Simo became well known for disrupting boat races—in one instance surfacing below the oars in a rowing competition and creating mayhem. His behavior toward people and general attitude were similar to those of Beaky and Percy in that he became excited and aggressive if surrounded by too many people. Like Beaky and Dolphy, he was fascinated by dogs—even favoring them above people when in the water. His home range was very small (only about ten square kilometers) by comparison with those of Beaky and Percy (up to seventy-seven square kilometers). However, Simo did not stay. In the winter of 1985–86, he disappeared. His solitary condition during two years was clearly a passing phase—a state that may be valid for all the solitary yet sociable dolphins that have been monitored.

DO SOLITARY AND SOCIABLE DOLPHINS NEED HUMAN CONTACT?

We can compare our dolphins' individual stories and behavior to find answers to the questions that we asked at the outset concerning their kind of contact and interactions with humans. Our studies of the three male dolphins in different life stages—juvenile, adult, and old, maybe very old—showed that each displayed many similar traits. Many behavioral patterns were also recorded for the five French dolphins—all of whom were female and also of different ages.

Following the stories of all the dolphins, it rapidly becomes apparent that each had a very different personality—just as we find for dogs, cats, or horses. There are very calm dolphins, like Jean-Louis off the Brittany coast, or calm and very distant and vigilant dolphins, like Marine who never engaged in any close contact with any person, or dolphins like Fanny, distant with most swimmers but willing to interact intensively with one person. Dolphy could not be described as "calm" or "distant" or choosing only some preferred interaction partners. Finally there was Françoise, who seemed unique (based on all known reports on solitary and sociable dolphins) because she shared her time both with boats and swimmers and with her natal dolphin group. She was also the only one of the French dolphins who did not disappear but remained in the lagoon together with another dolphin, though she ceased being sociable with humans. She drowned in a net in 2001.

The males also had different personalities but definitely showed more dominating, aggressive, and overtly sexual tendencies than the females, especially Percy, the old male.

Considering only our study of dolphins—without comparing them to all of the other known solitary and sociable dolphins in the world—we can already say that there is not one particular "type" or "kind" of dolphin that interacts with humans. And if we search for any definition of dolphin "sociability" with humans, we have to use a very broad one, including all sorts of interactions between the two species—as well as with domestic dogs. We must also consider dolphins that have regular contact with conspecifics and cannot therefore be considered truly "solitary."

Are solitary and sociable dolphins different from other wild dolphins, and is this the reason they become sociable with humans? None of the dolphins in our study had very close contacts with swimmers or divers at the beginning of their residence in a particular marine area. Most of these solitary dolphins had been discovered by local fishermen or other professionals of the sea, who were working in the area where the dolphin was present. These people usually reported that, at the beginning, the solitary animal was not particularly interested in interacting with people, being observed hunting and swimming around but eventually following boats inside the range.

At the moment of its discovery, the solitary dolphin can therefore be considered as *not sociable* with humans. But after getting used to human presence and activity in its sector, such a naturally inquisitive animal usually begins to show interest in the particular activities of the human users of its range and to have more confidence if some swimmers try to approach them.

The time up until the dolphin becomes sociable and allows itself to be touched may depend on how regularly and with how much patience one or several swimmers stay in the proximity of the dolphin in order to gain its confidence—as shown by the story of the very anxious Fanny and her friend Sylvia, who went to visit the animal every day. It certainly also depends on the individual history of every solitary animal: if, for example, it had had negative experiences with humans in the past. Several of our sociable dolphins had been exposed to human aggression—for example, Beaky, Fanny, Marine, Jean-Louis, and Dolphy.

Apart from personality, another factor in determining the time humans may need to gain the confidence of a wild solitary dolphin is its age. Dolphy was clearly a young juvenile or subadult when she first appeared in the Baie de Paulilles in southern France, as was Simo of Wales. Dolphy was reported to accept approaches from the employees of the aquaculture farm very rapidly after only several months of avoiding proximity. Simo accepted approaches by swimmers much more quickly than either the two old adult males Beaky and Percy.

The remaining question to answer is: do solitary and sociable dolphins *depend* on human contact? If we can agree that, at the beginning of their residency in a human neighborhood, "solitary and sociable" dolphins are just wild dolphins not yet sociable with humans, we must accept that when they appear in some particular marine area they do not need any human contact. Usually, as shown in nearly all known stories of such individuals, the solitary animals appear in a marine area where they have easy availability to food without needing other dolphins for cooperative hunting.

It is obvious that, at some level of habituation of the dolphins to swimmers, they become motivated in having more regular interactions with them. But does this behavior signify that the dolphins "depend" on human contact? Do they really need relations with people or one particular person, or do they just need social contact and entertainment? The family of Sylvia Chambon thought that Fanny was very "unhappy" when nobody came to swim with her in the channel. But can we say that the sociable dolphins are "unhappy" when nobody is swimming with them, or are they just bored? Boredom may explain why Fanny was swimming in circles when alone, in particular at the beginning. Dolphy also showed similar behavior patterns during long stormy days by circling around boats or buoys inside harbors. These behaviors are perhaps reminiscent of stereotyped behaviors in captivity.

The "playmates" of sociable dolphins may provide a kind of surrogate social partner. This was particularly evident in the case of Sylvia and Fanny, where one could say that the dolphin depended on the relationship with the young girl and her family. However, Fanny, just like most other solitary and sociable dolphins, left her range and her playmate Sylvia one day when more basic needs developed: she no longer found food in the channel and had to find another marine area to survive.

All of these examples show that the sociable dolphins seem to accept people as temporary substitutes for their conspecifics. Dolphy clearly preferred interactions with Rocky and other dogs to those with human swimmers, and she left the people immediately when a dog jumped into the water. Nevertheless, the sometimes intense interaction between dolphins and people, regardless of the dolphin's age and sex, is remarkable. Even after the human-dolphin interactions are analyzed in the most objective manner, something unpredictable remains that can only come from a blending of individual personalities and the chemistry of dolphin and man.

In closing, we must remember that, when we enter the water with dolphins, we enter their world and become subject to their rules and perceptions of what is acceptable behavior. The most important rule to remember is that all of the sociable solitary dolphins so far reported have often become uncontrollably excited and unpredictable when actively pursued, harassed, and crowded too zealously by too many people. We must recognize that, even if dolphins do not need human company for their social well-being, they often do need human protection.

NOTES

1. C. Lockyer, "Review of Incidents Involving Wild, Sociable Dolphins, Worldwide," in *The Bottlenose Dolphin*, ed. S. Leatherwood and R. R. Reeves (San Diego, Calif.: Academic Press, 1990).

2. M. Müller, "La place des dauphins solitaires et familiers dans la Socio-écologie

des Grands Dauphins *(Tursiops truncatus),*" Ph.D. diss., University of Paris VI, 1998, p. 437.

3. M. Ferrey, A. Collet, and C. Guinet, "Statut et comportement social du Grand Dauphin *Tursiops truncatus*, Mont. 1821 dans le Bassin d'Arcachon," *Revue Ecologie (Terre et Vie)* 48 (1993): 257–78.

4. Groupe de Recherche et d'Etude des Mammifères Marins de la Fédération des Sociétés pour l'Etude, la Protection et l'Aménagement de la Nature dans le Sud-Ouest.

RACHEL SMOLKER

Life at Monkey Mia

How and when dolphins and humans first made friends at Monkey Mia remains a mystery, but certainly it started long before we arrived. Some of the old-timers told stories about the dolphins who used to visit in the early days. Specifically, they remembered Old Charley, a dolphin who apparently used to come into the shallows during the 1950s. As the story goes, Old Charley would herd schools of herring into the shallows so that the fishermen could catch them easily, and as a reward for his effort, the fishermen would give him a few of the fish they caught. There were reportedly other dolphins that came into the shallows at that time, but Charley was the best known. I have my doubts about generous Old Charley. Even now the dolphins sometimes chase fish into the shallows at Monkey Mia, in proximity to people. Those who have witnessed this often claim that "the dolphin chased the fish in to me," or "the dolphin gave me the fish." I've witnessed this on a number of occasions, and to me it looks as though the exhausted and disoriented fish are seeking refuge in shallow water and is pursued there by the dolphin. There are some well-documented cases of wild dolphins hunting cooperatively with people (in Brazil, Mauritania, and along the east coast of Australia, for example). In Shark Bay, I suspect, the dolphins were not necessarily cooperating with people but just happened to be in proximity of the people who took advantage of the situation. . . .

An elderly woman by the name of Nin Watts claimed to be the first person to feed a dolphin by hand at Monkey Mia. She had stayed for some time on a sailboat moored just offshore from the campground in the mid-1960s and coaxed a dolphin to take a fish from her hand. I'm just not sure how she could know that she was the first, that nobody else had fed a dolphin at Monkey Mia before her, but that's what she claims.

I view all of these old accounts with skepticism, having seen how easy it is for dolphin identities to be confused and for stories or impressions to turn into "truths." Just as an example, I have watched countless visitors, having heard rumors that the dolphins could be distinguished by nicks on their dorsal fins, then assume that any dolphin with a nick was Nicky. And one old fisherman, a wizard with respect to fishing and fixing just about anything mechanical, used to tell us about seeing Holeyfin and Nicky at various locations all over Shark Bay, way beyond where those dolphins range, and at times when we knew they had been right here in the shallows at Monkey Mia. Somehow it hadn't occurred to him that there were many other dolphins in the bay. Anytime he saw a dolphin, no matter where, he assumed it had to be one of the Monkey Mia Dolphins. . . .

The true origins of human-dolphin interaction at Monkey Mia are unknowable at this point. Maybe the Aborigines started it, or maybe old Nin Watts really was the first to feed a dolphin at Monkey Mia. Maybe Old Charley was the original Monkey Mia dolphin, or maybe he was just the one who was around at that point in time which is now the limits of memory for the oldest of the old fishermen from whom we heard stories. With all the different stories and plausible scenarios, we can only guess that all may have contributed some part to building the relationship. My own guess is that the dolphin-human relationship at Monkey Mia was started by the local fishermen. Monkey Mia is within Shark Bay and is one of the only places where relatively deep water abuts the shore. The local fishermen have long taken advantage of this fact to bring their boats up to shore to clean out their nets. Just as they do now, they have probably always thrown a few fish to the dolphins while they were at it.

Interacting with people was apparently an acquired skill. An experienced dolphin approaches a person who has a bucket of fish and assumes the specific begging posture: pectoral fins braced against the bottom, head held up and out of the water, mouth open. Inexperienced dolphins seemed to find this awkward and would sometimes "practice." Joysfriend, a dolphin who became very familiar to us from our offshore observations, and who spent a lot of time with Holeyfin's daughter, Joy, did not usually come into the Monkey Mia shallows. But occasionally she visited with Joy and Nicky and Puck.

Once I watched her swim into shallow water at the shoreline, facing an empty beach. Without putting her face up, she stopped and looked around expectantly, as if to say, "Okay, so where's the fish?" She moved back offshore and seemed to watch the other dolphins approach people and take fish. A while later she tried again, this time bringing her face up out of the water but failing to open her jaws. She was going through some of the motions, but again, she was several yards from the nearest person and still didn't have her jaw open. She looked awkward, like a person taking his or her first stab at stage acting.

The habituated dolphins sometimes came into Monkey Mia accompanied by companions from offshore who normally did not come in. With a few interesting exceptions, these newcomers would usually hang back a bit. They weren't accustomed to contact with people and generally did not take fish handouts or permit physical contact. These rare visitors took an interest in what was going on at Monkey Mia, but they seemed to be at odds, both curious and shy. . . .

I once had the opportunity to feel what it is like to possess an animal curiosity. Several dolphins, a mix of both Monkey Mia regulars and a few who were unfamiliar to me from offshore, were chasing fish in the shallow water just offshore of the far west end of the campground. I waded out to watch them more closely and found myself standing in the middle of a feeding frenzy. A large school of fish swam around frenetically in the shallow water, and dolphins were tearing around me on all sides. Silver gulls and Caspian and Royal terns screeched and reeled overhead, diving down on the fish from above. In the shallow water, I could see the fish clearly and occasionally even caught a glimpse of a dolphin chasing and snapping up a fish in its jaws right alongside of me. After half an hour, the fish school dissipated and the frenzy died down.

Snubnose, who had been busily hunting up to this point, approached me now and put his face out of the water by my hand. I stroked him and spoke friendly gibberish. Then he put his head back under the water, turned to face offshore toward some other dolphins, and whistled very loudly. Two of these other dolphins, both of whom were strangers to me, were about forty yards out but turned at the whistle and came directly toward Snubnose and me. They came up alongside Snubnose, placing him between themselves and me. The two strangers seemed nervous. They were "holding hands"—

that is, they had their pectoral fins held against each other's sides—and their movements were fast and a bit jerky. The three dolphins circled me four or five times, eyeing me intently the entire time. Then the two swam off, leaving me with Snubnose. I felt like an exhibit in a zoo. These dolphins, apparently on Snubnose's invitation, had come over to have a close look at the human. And once they had taken a look, they swam off again.

Did I make an okay impression? Had they ever seen a human being before? What might they think about their companions, Snubnose and the other habituated dolphins? Did they think them brave for being unafraid of these aliens? Or dumb for allowing the aliens to touch them? Did they wish that they too could get fish handouts? Or did they view the Monkey Mia dolphins as social pariahs, lowering themselves to eating dead fish rather than catching their own? Did the Monkey Mia dolphins somehow keep others away from the campground area, defending it as their territory? Or was Monkey Mia considered the dolphin equivalent of Siberia? . . .

Meanwhile, we were not the only people interested in the dolphins of Monkey Mia. As word got out, more and more people rose to the challenge of getting out to Shark Bay to see the dolphins. With increasing numbers of visitors, tourism developers had dollar signs in their eyes while concerns were mounting about potential threats to the welfare of the dolphins.

Tensions mounted as we began to voice our dissent with respect to some of the dolphin-human interaction practices. In particular, we were concerned that people were selling little packets of "mulies," small, oily, anchovy-like fish, to the tourists to feed to the dolphins. The fish were frozen and were intended to be used as fishing bait. They often showed signs of decomposition and, besides, were not a species that occurred naturally in Shark Bay. The dolphins gobbled them down, even half-frozen. In fact, though they tended to turn up their rostrum at butterfish (fresh caught and local), they seemed otherwise pretty indiscriminate about what they would eat.

Once, a woman on the beach asked me what dolphins liked to eat. I explained that they preferred fish and probably also ate some crustaceans and other marine organisms. She pulled out a shrimp salad sandwich, replete with mayonnaise, and began dropping bits of the bread and salad onto the water surface, as if to attract fish. I explained that this might not work, that they only ate fresh, raw seafood, and that she would have better success if

she waited until a dolphin was actually present. A little later I came down
to the beach to find Nicky in. She swam up to me and regurgitated two
cooked shrimp along with a cloud of guck.

We worried that the dolphins would get sick. In their natural world, they
would probably never eat anything but the freshest fish—still alive, in fact,
and certainly not half-frozen or decaying. People argued that the dolphins
were smart enough to know better than to eat anything that might not be
good for them. But it seemed quite likely to us that, since they would not
normally be faced with frozen or rotten food in the wild, they might have
no evolved capacity to distinguish rotten from fresh food.

The sale of these packets of bait fish was turning into a major source of
income as more and more tourists arrived, seeking to have their picture taken
feeding the dolphins. We suspected that this financial concern might be at
the bottom of the resistance we faced in trying to change dolphin-feeding
policies: the dolphins were being stuffed full of junk food.

We also worried that someone might hurt the dolphins accidentally or
even intentionally. In other places where people and dolphins had taken to
interacting, the dolphins had often enough ended up dead. One such case
occurred in Opononi, a tiny, sleepy fishing community in New Zealand,
where a young child had befriended a dolphin, word got out, tourists began
flocking to town, and not long afterward the dolphin was found dead. Word
about the Monkey Mia dolphins was already spreading far and wide. The
road into Shark Bay would soon be paved all the way from the Overlander
to Monkey Mia. Many more people would be able to make the journey. This
isolated, lazy little place was poised on the edge of radical changes, and there
were many conflicting opinions about how to progress.

One thing everyone (the local government, the tourism interests, and
ourselves) agreed upon was the need for some presence on the beach to monitor human-dolphin interactions. Someone needed to make sure that nobody did anything to harm the dolphins and also to field questions that
visitors might have about the dolphins.

A small trailer was donated by the "Golden Dolphin Tribe," a band of
hippies who had made a pilgrimage from Sydney across Australia to visit
the "mystical" Monkey Mia dolphins. The trailer was marvelously decorated with a painting of dolphins leaping in front of Ayers Rock (which lies
dead smack in the center of Australia: as far from ocean as one can get on

the continent). Richard and I were encouraged to avail ourselves to the tourists to answer questions about dolphins, to help keep an eye on things, and to aid in developing the trailer as an interpretive center. . . .

What would dolphins talk about if they could talk to us? Would they discuss the latest good fishing spots or tell stories about their encounters with sharks? What if we could ask them questions like: Are you really as smart as we think you might be? Can you stun fish with your clicks? What do you do with those sponges? What could they tell us about themselves and about their ocean world? . . .

I and other researchers have put many, often tedious hours into analyzing dolphin communication, always with the hope of "breaking the code" or making some key discovery that will allow us finally to pull back the curtains and understand these animals. But I have also wondered what the dolphins think of *our* communication. They hear us speaking constantly. Do they recognize that there is something special about that? The closest I have come to an answer to that question came one day when I was swimming with Nicky. It was always a challenge to engage her attention in a good swim. I suppose she knew I wasn't likely to have a fish handout, and I had the impression that she found the flailing swimming motions of humans irritating. While Puck was usually game and would often approach, whistling, and swim in circles around me or allow me to put a hand against her side, Nicky almost invariably ignored me. But on this occasion I was able to really capture her attention. I got into the water and she gave me an "Oh, it's just you" sort of look and headed away. I searched my brain for some inspiration, something really different and interesting that I could do to interest her. On the bottom just below me I noticed a stick, about a foot long and perfectly straight. I dove down and picked it up, not sure what I would do with it. The feel of it in my hand gave me an idea. I started to write with it on the sandy bottom "I LOVE NICKY" in big stick letters.

Nicky turned, circled me at a bit of a distance, and then moved closer. She came right alongside me and put her face up to the tip of the stick, her head moving in synchrony with the motion of my hand. Her attention was riveted as she followed every movement intently. I wrote my message to her, moved over to a fresh patch of sand, and wrote it again and then once more. She followed me each time. When I stopped several minutes later, her eye rolled to meet mine, then back to the writing in the sand; she paused and only then swam away. I had never before, and have never since, held her attention so completely.

I was dumbfounded—not just because I had won Nicky's attention, but because she seemed to realize that this behavior of mine, writing, was something significant. Something worthy of her attention. I don't begin to imagine she understood the meaning of my message to her or that she truly understands how profoundly important writing is to the lives and cultures and communication of humans. But her extraordinary attentiveness to my stick writing made me think that she recognized at least that something truly important was going on. That is exactly how I feel about the communication of dolphins and why I chose this difficult topic as the focus of my research; it is difficult if not impossible to understand, but there is definitely something profoundly interesting going on. . . .

As more and more people came to visit the dolphins, the interaction became, by necessity, more and more closely managed. A team of rangers had been hired to ensure that nobody, dolphin or human, got hurt and to answer visitors' questions. This was critical. Things had gotten chaotic as large crowds of tourists scrambled to touch and feed the dolphins.

Watching the interactions of people and dolphins was an ongoing source of both insight and entertainment for us, ranging from the sublime to the ridiculous to the downright dangerous. For some people, interacting with a dolphin is a highly emotionally charged affair. I know some of my own thoughts and feelings about interacting with the dolphins went way beyond what would transpire in an interaction with the neighbor's dog. I remember some of my feelings during one of my first encounters with Nicky. I was standing on shore, and Nicky was floating just five feet from me, on her side. One eye was tilted up and watching me. I waded into the water. "Nicky! Good morning, gorgeous. Are you waiting for something?" Her eye followed my hand, then rolled up to look me in the face, back to my hand. I reached toward her to stroke her side, and her jaw opened just a little bit, then stopped in a slightly open position, revealing her perfect row of teeth. She must have thought I had a fish offering, then realized it was just my hand. Her body drifted away from me a bit, and I took a step closer. She was watching someone else walking on the beach. I could imagine her thinking, Does that one have any fish? This one is clearly a loser. Just wants to paw at me with those dangly mitts.

Her tail flukes bumped my shins as she moved off to approach the new-comer, a gangly boy of about fifteen. The kid was tentative. He looked at her but stayed back, holding his arms stiffly at his sides. His demeanor revealed a combination of amazement and fear. Nicky rolled on her side and watched him. We all stayed put and waited. Holeyfin surfaced about fifty feet off, foraging. She took a couple of breaths, and then her tail came out as she dove. Nicky suddenly turned out toward her, as if she had heard something interesting, but then stayed with us.

It was a peaceful scene on the beach, just two humans and a dolphin, quietly enjoying the moment. But I was quick to become restless. I wanted Nicky's attention. I wanted to play with her, to touch her, to explore her responses. And if she responded badly to me, I would be embarrassed and mortified. I wondered if the kid felt the same way. Maybe part of his fear had to do not so much with concern about getting hurt as with whether or not Nicky would like him. What could be worse than being disliked by the most altruistic and loving, perpetually smiling and happy creature of all? I approached Nicky again, and she swirled her whole body around in order to see me. I spoke to her, saying ridiculous things intended to convey my friendly intentions. She stayed put as I came close and paused, then reached out to stroke her. She suddenly jerked her head out of the water toward me, clearly agitated. I was heartbroken. I backed up out of the water, chastened, and stood still, telling myself that if I had a fish, she would have been nicer to me. Maybe she just didn't like the color of my purple shirt.

The kid decided to have a go. He approached Nicky, and she moved closer and poked her rostrum against his knees, rolled her eye up at him, and emitted that questioning little *eheh?* vocalization that the dolphins always make when they are asking for fish. There was no sign of the irritability she had shown toward me a moment ago. She allowed him to stroke her side, leaning against his thighs, then nibbled at his toes and moved away. I was intensely jealous.

For many visitors, perhaps even the majority, the goal is to get a photograph taken of themselves feeding a fish to a dolphin. Wading in tentatively, fish in hand, they dangle the fish just in front and out of reach of the dolphin while a partner focuses the camera. Once the photo has been taken, it's time to hit the road. "Been there, done that, checked it off the list." In recent years, tour bus operators have begun to carry loads of people up from Perth. After the long trip, the passengers are discharged onto the beach for an hour or two, plenty of time for that photo opportunity and a visit to the snack bar. Then they load up for the ten-hour journey back to Perth.

I feel sorry for these folks. They have made such a long trip out to Shark Bay, and their experience seems so impoverished. I imagine they must feel gypped, but perhaps it is all they expected. Once I watched a chap pull up into the parking lot in his land cruiser, roll down the window and assess the situation, then drive directly down onto the beach in front of the dolphins, snap a picture through his car window, and drive away.

What disturbs me most about these drop-in tourists is that they don't seem to realize just what an incredible privilege they are being granted. Through their eyes the dolphins are reduced to quaint little critters, put here to provide a trifling amusement and a photo op. . . .

There are many visitors for whom seeing, touching, and feeding a wild dolphin is the fulfillment of a lifelong dream. One incident among many stands out in my memory: An old couple standing with just their toes in the water. The woman's eyes have the pale and disarming look of a blind person, and she is holding her husband's arm. Her face is positively brilliant with excitement. Her husband very gently helps her into the water, instructing her to pull up her pant legs. She is so excited that she can hardly stand to be bothered and starts to lose her balance a bit. He steadies her with his shoulder and leads her forward toward the dolphin. She is reaching out toward the general direction of Puck, fingers pumping, seeking. Her face is quivering with excitement; her eyebrows, raised high with elation and expectation, smooth some of the many wrinkles on her aged face. The emotion is spelled out on her face, uncensored in the way of the blind. Hand groping, she says to her husband, "Oh, is there really a dolphin? I want to touch it. Lead me close so I can touch it."

Everyone else stands back in deference to this couple, clearing the way so they can approach Puck. The blind woman is still leaning against her husband, who is completely and gently attentive to her. She leans forward and reaches out. Her husband takes her hand and carefully guides it to Puck's side. The woman's face registers a thousand feelings in that instant. She is surprised by the warmth. Her hand, filling in for her lack of vision, explores Puck's form, and Puck, seemingly aware that this is a special occasion, tolerates the groping. The husband talks to his blind old wife, holds her arms, directs her hand, then after a time leads her gently back to shore. Her pants are soaking wet. There are tears rolling down my face.

The Monkey Mia dolphins are experts in human behavior. They have seen it all, but they still take an interest in anything out of the ordinary. I have seen them examine at great length pregnant women, babies, people on crutches or in wheelchairs, or people who just plain look weird.

One day I watched Nicky exploring the wonders of human anatomy. It was after a feeding session on a calm day with a small group of easygoing tourists. Nicky and Puck were swimming slowly back and forth among the people, some of whom had drifted out into waist-deep water to cool down. Puck had approached a middle-aged woman and was "parked" in front of her, mouth agape, face out, lingering there and allowing the woman to stroke her side. Meanwhile Nicky had circled behind the woman, who was bending over toward Puck. Nicky stopped right behind her, staring intently at the woman's bottom, turning her head to look first with one eye and then the other. It was, in fact, an amazing bottom: tremendous, with lumps and rolls and bulges overflowing from a too tight bathing suit. Nicky continued her examination, putting her face up to within a few inches of the woman's bottom and growing wide-eyed as the woman moved slightly, reaching farther toward Puck. The woman was unaware of the attention she was receiving from behind, even when Nicky reached out her pectoral fin and stroked the woman's bottom several times. Nicky decided to investigate further. She moved even closer, poking her rostrum into one bulging cheek and then backing off to inspect once again. I could imagine her thinking, Wow, look at this! What does it feel like? The woman remained oblivious through it all. . . .

We had come a long way toward making sense of the lives of the Shark Bay dolphins. The more we had learned, the more discoveries we realized still lay ahead. But personally I felt it was time for my own life to take precedence. It was time to settle and put down some roots, cultivate relationships that could endure, finish up the many reports that lay half-completed on my desktop. It was time to slow down on my visits back and forth to Australia and take time to reflect upon my experiences and all that we had learned from the dolphins. . . .

Still, when I shut my eyes, I hear the ring of the wedgebill's song, and out of the darkness emerges a dolphin, looking up at me through the water surface, that transparent but profound barrier between us. From such different worlds have we come that we could as easily be aliens from another planet. Yet there is so much that we share. For one thing, we are both curious enough to overcome any fear that might have prevented us from even meeting.

What are you? Who are you? For both of us, the answers to those questions lie in long histories—both evolutionary and personal. Our lives have twisted and turned from birth, through many years of childhood and on into long adulthoods, encompassing great transformations, pleasures and frustrations, good fortune and loss. Though our species bear no resemblance, we have both learned to navigate the tricky waters of our relationships with others, loving, loathing, giving, and taking, trying always to understand what makes our friends and associates tick. We both carry these long and intricate histories into this moment with a common interest. Can we be friends? Hey, you, on the other side there, our hearts and minds are not so different after all!

Miracle Dolphin

For hundreds of years, the nomadic Bedouin people of the M'zeina tribe have lived along the shores of the Red Sea's Gulf of Aqaba on the east coast of Egypt's Sinai Peninsula. They have fished the deep, clear, steep-shored gulf waters and have held fiercely to their traditional ways, including a reverence for the sea and all its living creatures. These "sea people" hold a particularly deep respect for bottlenose dolphins, known in Arabic as *Abusalaam* or "the father of peace," with whom they have traditionally shared their fishing grounds.

In the spring of 1994, I heard a rumor that a "lone sociable dolphin" was interacting with the local Bedouin fishermen in the village of Nuweiba M'zeina. By definition, a "lone sociable dolphin" is an extraordinary phenomenon. As a member of a toothed-whale family, this dolphin changes its regular life habits from membership in a pod of fellow dolphins to a voluntary association with a completely different, terrestrial, companion species—*Homo sapiens*. The reason these dolphins leave their own pods of their own free will and make contact, at times very close and long lasting, with humans is still unclear. But I was intrigued by this mystery after an exceptional encounter I had had with another lone dolphin two years earlier.

In the summer of 1992, in Eilat, at the northernmost point of the Gulf of Aqaba—on the border between Israel and Jordan—I met "Crispy," an adult male bottlenose dolphin. Crispy was a lone, sociable dolphin that for a short period of a few months interacted with humans on a daily basis. Crispy used to visit the commercial sea fish farm, which was located right on the border, on the north shore of Eilat.

From the fish farm workers, I learned that during their cage-maintenance dives the dolphin used to escort them and at times even "carried" and hid equipment items such as hammers and other working tools. His favorite game was changing locations and surprising the swimmers each time from a different direction. Only after a few months of observation and staying around divers who were constantly working in the area did the dolphin allow familiar divers and fishermen to touch him. Soon, touch was solicited. He faced divers in a vertical, belly-to-belly position so they could stroke and scratch him. When the petting stopped, Crispy would bite the person gently, as if asking for more.

Crispy developed at least one close relationship with a spear-gun diver who used to fish in that area. The diver claimed that Crispy would momentarily shock or disorient fish by emitting bursts of intense, high-pitched sound. All that was left for the diver was to shoot the stunned fish and collect them.

Then, at the end of December 1992, after observing Crispy for several months, I had an encounter that to this day I cannot explain. My diving buddy at that time was my good friend Aharon David. Toward the end of the dive, while starting the ascent, we realized that we had forgotten the underwater video camera on the bottom of the sea. When we started our search for it, I did not notice that my tank was very low on air. We found the camera and I started to swim slowly to the surface with Crispy escorting me all the way up. At the depth of nine feet, I insisted on performing a safety decompression stop. But by now I was completely out of air and my desperate inhalations emptied my mask and started squeezing my eyes. In those critical seconds, Crispy swam underneath me and forcefully pushed me to the surface.

This was the first time in our brief acquaintance that Crispy was so persistent, and I believe that he somehow sensed my difficulties in breathing and decided to help me out. To this day, I do not know what led Crispy to assist me, since I could not be considered his regular pod member.

But my story is not unique, and such stories of wild dolphins assisting swimmers in distress in first-time encounters abound. The news of the lone dolphin in Sinai was thrilling. If the story turned out to be more than a rumor, this less-accessible animal could be a golden opportunity for me, a "private" dolphin for my research. In June 1994, I arrived in the village of Nuweiba M'zeina. I was surprised to see that the local Bedouin fishermen were still swimming and snorkeling without masks and fins. A brief investigation in Arabic with the Bedouins informed me that there was indeed a lone dolphin, but they did not know its species or gender. Since the beginning of May, the dolphin had been coming to the area on an irregular basis. It would return to the area every few days and escort the Bedouin dinghies on their way back to the village. Sometimes it even lingered off the village shore for hours before disappearing from view.

I had no way of knowing when or if I would ever see this dolphin, but, luckily, after I had spent only a few hours in the village, the dolphin arrived. Some of the local Bedouins ran into the water and started swimming with it. I grabbed my snorkeling equipment and jumped into the water. I joined the Bedouins who were swimming beside the dolphin and recognized her to be an adult, female, Indian Ocean bottlenose dolphin. An unknown European tourist called her "Holly."

The Bedouins tried to touch the dolphin, but Holly would not permit real physical contact, although she let them come to within a few centimeters of her body. In the first few weeks of her appearance in the area, she swam side by side with her new human companions and seemed to build up the trust and bonds with the locals. I knew from my experience with Crispy that this was actually the first stage in the relationship between a lone, sociable dolphin and humans.

During those early days, I also came to realize that many of the Bedouin fishermen swimmers were deaf. The word in Arabic for deaf is *atrash*. Because of their condition, they were considered lower on the social scale and could not find any "respectful" occupation. So they remained fishermen and became the first ones to swim with Holly, spending hours in the water. Among the group of Bedouins at the time, there were two people who were outstanding in their zeal for forming a relationship with the dolphin: Muhammad Salaam Atwa, deaf and mute from birth and nicknamed Muhammad El-Atrash; and his cousin Abd'allah Mehaisen.

On July 1, after Holly had been in the area only a few short weeks, I watched as she "spy-hopped," a social behavior where the animal lifts its head vertically out of the water with eyes above the water, with beak pointing skyward. She was taking a peek at the local swimmers who sat on the beach. Later that afternoon, an English tourist who was staying in the village came out of the water and almost fainted on the beach. I approached her and, when she came to, she told me very excitedly that she had touched the dolphin. Only then did I realize that the ice was broken and Holly had begun to trust humans. I entered the water and saw that I was right. Holly swam close to me and positioned herself vertically in front of me and moved up and down. With her head out of the water, she allowed me to touch her. I had been in the Bedouin village for only a few weeks and I had actually touched the lone, sociable dolphin!

I knew that I wanted to study Holly and her interactions with humans. I also knew that, in order to begin my research in this village, I would need to earn the trust of the villagers. In the Bedouin culture, honor is a very important code, and, once you acquire it, you are assured of a lot of respect, which is equivalent to support and defense when it is needed. As I came out of the water, I noticed Abd'allah Mehaisen. "Come into the water with me, the dolphin is allowing me to touch her," I explained.

"No one touches my dolphin," Abd'allah, who is a very spirited individual, partly deaf and hard of speech, yelled back to me. Then I realized that, if I could convince him that I knew what I was talking about, I would earn his trust. I understood that it was the only way, if I wanted to start my research in the village.

It took me a while to convince Abd'allah, but finally he agreed to enter the water with me. A moment later he touched the dolphin. I could sense his excitement. When he got out of the water, he yelled to the Bedouin who sat on the wall, "It's true. The dolphin is touchable."

The Bedouins accepted me as part of the family and as a self-proclaimed authority on dolphins. I could start my research. From June 1994 through July 2001, I documented the development of Holly's long-lasting human relationship.

Through the first year of her stay among humans, Holly was extremely docile and seemed pleased with human contact while she was swimming around the fishing village. Unfortunately, some people took advantage of this situation and abused her by harassment, disturbing her when she was resting and petting her when she did not want to be touched. She remained passive and even allowed physical touch that included mouth grabbing, forceful spreading of the jaws, and forceful spinning and rolling by humans.

Gradually, she began putting her human pod members in their place as human misbehaviors became more prevalent. She usually gave "fair warning" signs before inflicting damage. In most of the cases the swimmers did not correctly interpret the body language of the dolphin. When the swimmers misinterpreted the warning signs, Holly corrected them but never too forcefully. In most cases she bit their fingers slightly, slapped them with her tail, or butted them with her head.

During my seven years of research, Holly became pregnant and gave birth three times, twice to male calves and once to a female calf. Holly raised all her calves among humans. Unfortunately, Holly lost both male calves. There can be negative impacts for a calf that is raised among humans, especially when there are almost no limits on its interactions with them. With her third calf, Holly changed her behavior: she became more protective of the calf by preventing her from intensive interaction with humans; she also extended her daily home range. As of October 2000, Holly stopped coming to the M'zeina village on a daily basis to interact with humans. However, the fact that Holly raised her offspring among humans made her an extremely unique dolphin.

Over the years, Holly became the leader of the mixed dolphin-human pod. In many cases, I observed Holly while she, of her own free will, swam toward familiar "human pod members" and, upon arrival, leaned on them. In the wild, a dolphin swims above another individual and puts its pectoral fins on top of the other's back as a signal of dominance. It may be that Holly sees herself in a dominant position in her substitute pod.

Eventually the word spread, first in Israel and Egypt and later in Europe, and a "swim with the dolphin" tourism, specifically or as a part of a Sinai touring package, started to flourish and has been increasing annually. Swimming fees were collected, snorkeling equipment was rented, bungalows and cafés were opened. Busloads of tourists, at times arriving simultaneously, caused situations with scores of swimmers in the water at any one time.

Lately, as a general rule, when swimmer pressure becomes excessive, rather than drastically modify her behavior or swim just out of reach, Holly will elect to move into secondary "home ranges" on less crowded beaches, returning early in the morning of the next day. I believe that she returns not for food but for companionship. There were times when the Bedouins offered Holly a small amount of food, which basically was not more than 5 percent of the daily food that a dolphin of that size requires. Holly only accepted live octopus, handed to her by one of the Bedouins. On many occasions, Holly treated this offering as a game rather than as a piece of food.

After almost seven years of research, I believe that, despite the physiological differences between humans and dolphins, a long-term social relationship largely based on the dolphin's own behavioral language is a preferable alternative to a truly solitary existence for a highly social animal like the dolphin. In order to encourage and nurture the relationship, the human companions, even in single encounters, should be as aware as possible of that dolphin language.

Holly's presence has had an enormous impact on the Bedouins. The whole village has changed and become a major tourist attraction in Sinai. Many young Bedouins have spent their youth growing up with the "miracle" story that exposed them to the world of the dolphin. I have no doubt that all the people who have interacted with this amazing dolphin can be the best ambassadors to start a new wave to preserve the ocean and the magnificent creatures that live within it.

ERICH HOYT

Toward a New Ethic for
Watching Dolphins and Whales

In June 1998, I traveled to northern Japan to speak at a whale-watch conference on the subject of how best to observe whales and dolphins. Over the past decade, since whale-watching began to take off in Japan in the early 1990s, I have participated in four such forums at locales all over the country. These conferences make for warm gatherings of Japanese and international scientists, operators, and enthusiasts—a place where they can shake hands or bow low, exchange ideas and stories, and speak about amazing cetacean encounters.

Minutes after my long, overnight flight over Siberia had landed, the chief conference organizer, Dr. Tadao Furuya, a dentist who runs the local whale-watch tours, drove me into Muroran, on mountainous Hokkaido Island, where the conference was being held. We boarded a modern, well-equipped boat decorated with whale motifs and set off. Passing under the majestic Swan Bridge that acts as the gateway to the harbor, we found light breezes on twenty-five-mile-wide Volcano Bay, with scattered clouds above. The wind on my face eased the jet lag, and I felt the way I often do setting out to sea on an adventure to find cetaceans: anything could happen; it might be the chance to witness some rare behavior—a once-in-a-lifetime experience.

Onboard the fifty-foot boat were an American and a British scientist and his partner, as well as about fifteen Japanese, including several dolphin enthusiasts with a family of first-timers, plus photographers, journalists, two scientists, a cartoonist, an environmental campaigner, and of course the skipper of the boat. For such an assortment of people, things seemed a little formal. "Konnichiwa!" I tried to break the ice with my Japanese "hi," which predictably brought smiles. Slowing down my English, I learned that my companions had traveled from all over Japan for the same shared purpose as those of us from abroad: to seek out experiences with cetaceans and the people who like to watch them. We were in the right place!

After an hour of combing the rugged coastline of the bay, we met the local Dall's porpoises—about two dozen of them. They raced around our boat, creating characteristic rooster tails in the water and defying even the most agile Japanese photographers' attempts to get a shot. More than once, the photographers smirked as they bumped into each other, and then they became convulsed with nervous laughter as one became entangled in another's camera case straps.

The hyperactive Dall's porpoises, the largest of the five true porpoise species, are found only in the colder waters of the North Pacific. I had often seen them on my summer expeditions with orcas off northern Vancouver Island in the 1970s. They used to dance around the orcas, who moved slowly and steadily compared to the Dall's; we always joked that the porpoises, who sometimes become dinner for marine-mammal-eating orcas, have to keep moving. But certainly there are other, more likely reasons for speedy maneuvers—such as the animals' metabolism, the frigid water, and the speed of the fish they hunt. Their speed does not keep them from being killed by the tens of thousands in high seas drift nets in the northern North Pacific.

Dall's porpoises are mostly overlooked and are not the sort of cetaceans that engender public notice, much less support. Yet our little group of watchers loved their every move. I was impressed with the skill and enthusiasm of the Japanese skipper—my research and writing on and passion for whale-watching has enabled me to see and compare whale-watching going back three decades from all over the world. Being on a whale-watch boat in Japan with Japanese tourists—and 95 percent of all cetacean watchers in Japan are Japanese—is a bit like being on a California gray whale watch ca. 1972. There is raw enthusiasm, sheer exuberance. Fun. The oohs, aahs, wows erupt, and smiles are in abundance. The Japanese do not hang their emotions out in public as do Americans, but whales and dolphins can make them share their feelings faster than a few tumblers of sake.

Besides Dall's porpoises, we hoped to see Pacific white-sided dolphins, maybe pilot whales, and even orcas. Minke whales, the smallest baleen whale, are also found in this area, but we did not need to encounter them to make this a successful trip. Japanese cetacean watchers, especially young women, are fond of dolphins. Even on sperm whale–watching trips out of Shikoku Island and near Taiji, on Honshu, the main island, Japanese women often tell the skippers: "Please take us to see dolphins."

After leaving the porpoises, our eyes and ears were sharpened toward unusual splashes or puffs. As the day brightened, we soon spotted a herd of at least 200 Pacific white-sided dolphins. We began traveling with them, staying off to the side and moving parallel. After chunky Dall's porpoises, Pacific white-sided dolphins look lithe, streamlined, with delicate painted faces and flanks. They are more athletic, and now the oohs and aahs reached new crescendos as a small group of dolphins came to bow ride and leap beside the boat. Even the British, American, and Japanese scientists among us joined the cheers. For nearly two hours, we stayed with them, hungry for more close encounters. We wanted it to last. And then the rare and extraordinary happened.

About 200 feet off the bow, we noticed that two mature dolphins were assisting a calf, supporting it so that it could breathe. Other dolphins were circling around the three.

"I think it's a newborn calf," said one Japanese scientist. "Let's find out."

"It looks to me like an injured calf," said the British woman.

"The other dolphins are lifting it up to breathe," said the scientist. "We need to get closer."

At least ten cameras were poised at the bow. Many wanted to move closer. But not everyone: "Leave them alone!" shouted the British woman. "It's obviously a mother and calf and the mother knows what to do."

"If it's a calf in trouble," said the American scientist, "we might be able to help." He hesitated, then added: "Or we might make things worse."

"We should rescue the calf," said a Japanese diver dressed in his dry suit and ready to go.

We were only ten miles away from a village that kills dolphins with electric lances. It seemed that we should not be arguing over how to watch dolphins. Any way of watching dolphins with nonlethal intentions deserved our support in these waters. The "thirst for the close encounter" is, after all, human nature.

Still, several Japanese dolphin watchers sided with the British woman, who repeated her fervent wish to leave the dolphins alone. Then the Japanese skipper of the boat silenced the debate.

"Let's see what's happening," he said. And so we slowly motored over toward the calf and its entourage. As we approached, however, the surrounding dolphins dispersed, one by one, until only one adult and the calf remained. The British woman was upset.

The Japanese diver, also a photographer, yelled to the captain, "Stay here! I'll jump in and swim over."

Most of us—scientists, conservationists, and Japanese dolphin watchers—applauded this "compromise," though we were already fairly close. We studied the diver as he swam ever closer to the dolphin. Were the dolphins disturbed by the diver? Probably. When the diver was within thirty-five feet, the last remaining adult dolphin swam off. Was that the calf's mother? We did not know. The British woman was quiet and red-faced—anger tinged with embarrassment. Some of the Japanese were embarrassed, too. And the scientists were starting to question the suggestion to intervene.

Minutes later, the diver cradled the baby dolphin in his arms and swam back to the boat. He was holding it like a dead child. He passed it up to the waiting scientists. We crowded around for a closer look. It was a calf, a few hours to a few days old, and it had died, at least some hours before. Looking in its mouth, we glimpsed the tiny, perfectly formed teeth. It was like looking at a human baby's fingernails—all the delicate features of unrealized maturity visible in miniature. It turned the joyous afternoon into a somber occasion. As the photographers snapped away, there were a few tears. Then, a scientist from a local university said they would do a necropsy, and that seemed to cheer most of us; at least the cause of death might be found and the carcass might help other dolphins.

The event was memorable, not just for witnessing the dolphins who had stayed by their dead pod-mate, trying to keep it breathing, but for seeing the culturally based reactions and behavior of the people on the boat. The baby dolphin had almost certainly died before we had seen it or decided to come close. But it helped crystallize for me the issue of how we should be watching whales and dolphins—the subject of my talk, which would close the conference at week's end.

Was it appropriate to disturb the behavior of the dolphins who were attending the baby—even when we feared something was wrong? We know nothing about how these social animals might deal with the loss of family members or even "mourn" their dead. That baby's mother, aunties, maybe sisters were in attendance doing a "duty" that was arguably a mix of instinct and culture. To what extent did we intrude? And the larger question is whether humans engaged in watching cetaceans should seek to interact with them or influence their behavior at all. The ethic of watching wildlife without disturbing it, which has been accepted with many land mammals and birds in national parks and reserves, does not seem to apply to cetaceans in the same way. Even though whale-watching guidelines and regulations have been established in many parts of the world, few rules prevent interaction, much less disturbance; most simply try to regulate close encounters with mixed success by cautioning boat captains against making direct approaches and insisting they maintain a certain arbitrary distance. Yet even if these basic rules are followed—often they are not—cetaceans can hear boats coming from miles away. They know they are being watched.

The idea of interacting with dolphins reached the public consciousness in the late 1960s with the late John C. Lilly's interactive experiments with cetaceans. If he did not start it, he certainly pushed the concept along and popularized it. Lilly looked at the large, convoluted brain of the dolphin and wondered what it could be used for. He set up a long-term, live-in experience with a woman and a dolphin, the woman taking up residence beside the dolphin's pool. He famously stated that humans were on the brink of communicating with dolphins and gave a date, long since passed, when the magic breakthrough would come. Lilly's work was conducted exclusively in captivity, but it had an influence on many wild cetacean researchers.

Yet even if Lilly had not influenced a generation of cetacean scientists and enthusiasts, a unique aspect of cetaceans would have put the interaction question on the agenda sooner or later: certain cetaceans regularly approach humans. Most dolphin species visit boats to ride the bow or the wake or just to investigate. Some species, both the larger cetaceans such as humpbacks and gray whales, as well as orcas and certain dolphins, have so-called "friendly" individuals who often approach close to whale-watch boats. On top of this, several dolphin species, led by bottlenose dolphins, have the phenomenon of sometimes becoming "lone, sociable dolphins" eager to interact with humans.

At times, it would seem, it is all we can do to avoid cetaceans. Yet close encounters probably represent less than 1 percent of all cetacean encounters. Still, such encounters have made many operators and prospective whale watchers seek out the close encounter, to feel that it is all right to encourage or take advantage of cetaceans' accessibility and the habits of bow-riding and social or predator curiosity. Commercial trips often promise close encounters, and the trips are sold on this basis. The brochure photographs and the film and video close encounters all hint that close approaches and interactions confer the ultimate experience.

It is not hard to understand why. I will never forget one of my first close encounters with an orca in a small boat off northern Vancouver Island, British Columbia, in 1973. As I wrote at the time, he blew and his breath covered me like a cool shower on a hot day—and I was sweating. I believe it was partly instinctive predator curiosity and partly a playful sense that led the orca to inspect me, lifting his head out of the water from only a few feet away. Then he turned his eye on me. This animal proved to be half of a duo, a couple of youngsters we called the Twins, who, over time, often approached to swim around our boat and even ride the bow and the wake, behavior not normally seen in orcas. We eventually came to seek out the close encounter, hungry for the thrill of the unexpected. It was easy to get them going. A quick buzz in our Zodiac inflatable around the bay often did the trick.

This "relationship" with the Twins lasted two and a half summers. They soon matured and lost interest in play. Whether we were a negative influence on their adolescence, diminished their ability to survive, or lowered their "quality of life," is impossible to determine, but doubtful. We were only a few researchers and film-makers in the 1970s. Back then, there was no whale-watch industry for orcas. Worldwide, by the mid-1970s, there were only a couple hundred thousand whale watchers a year, mainly looking at gray whales in California and humpback whales in New England. Dolphin watchers, before the accessible spotted dolphins were discovered off the Bahamas in the late 1970s, were mainly confined to opportunistic sightings on California and New England whale watches. I suppose we felt that it was all right for ourselves and a few others to have close relationships with orcas but not for the general public. Yet something happened that changed that view and drove us to publicize our experiences with orcas.

In 1980, the largest logging company in Canada at the time, MacMillan Bloedel Ltd., announced that it was going to be logging the last untouched river valley on eastern Vancouver Island, driving the logs down the virgin Tsitika River and booming them in the bay at Robson Bight, which orca researcher Michael Bigg had identified as the core area for orcas. In this area we had often encountered orcas; here, pods came to rest, socialize, and rub themselves on smooth pebbles found on several beaches.

Through a poster and media campaign, we—a small ad hoc group of grassroots conservationists—put orcas on the agenda in the province of British Columbia as well as across Canada, and the ultimate result was that a marine protected area was created to help conserve the area for the orcas. It did not stop the logging of the valley, but it prevented logging in the estuary, including log booms and logging traffic, and the immediate disastrous effects all this would have had on the orcas.

The other result was that, after high-profile articles in *National Geographic, Equinox, Defenders,* and other publications, as well as my book, *Orca: The Whale Called Killer,* orcas were now in demand to be watched. The first trips in the early 1980s led to powerful word-of-mouth, and by the late 1980s orca-watching had spread from northern to southern Vancouver Island and into Greater Puget Sound. The numbers today, post-Keiko and the *Free Willy* films, amount to more than 75 operators taking 472,000 orca watchers a year in the northwestern United States and Canada alone. This may well be too many orca watchers, especially if all those people expect close encounters. Fortunately, more than half of all orca watchers, some 265,000 people a year, are watching from land-based spots such as Lime Kiln Park on San Juan Island, Washington. That helps cut down on the numbers of boats but still leaves a lot of close-encounter seekers on the water.

The "great orca trade-off," as I call it, resembles the trade-off for most other cetaceans. Many whale and dolphin species have particular habitat, hunting, or other pressures that have been eased, in one way or another, following the spread of whale-watching over the past few decades. The public is now more aware of cetacean problems and is willing to help in various ways through campaigning, assisting with research, and supporting conservation. At the same time, several thousand businesses in some 500 communities worldwide are supported by whale and dolphin watching. Cetaceans are so appealing that, according to my research, whale-watch tours grew through the 1990s at a 12 percent rate of increase per year— three times the growth rate of all other tourism. As of 1998, more than nine million people a year were going whale- and dolphin-watching, spending more than $1 billion. Since then, in the post-9/11 world, whale-watching has continued to be popular, though growth has slowed along with most other tourism.

Clearly, whale-watching is vital economically and acts as a sales tool for cetaceans and cetacean conservation. But it is time to reconsider the best way for masses of people to watch cetaceans. Is it with close encounters? What worked with whale-watching in the early years, when there were a few hundred thousand spread-out whale watchers, does not work as well now. There are not enough accessible, friendly cetaceans with time available—after feeding and socializing—to allow for nine million close encounters per year. Of course, in some parts of the world whale-watching is new, and the promise of closeness sparks young businesses. But the new operators need to look at what is happening in the busy, established whale-watch capitals to know what to avoid: too many people, too many boats on the water, all trying to get too close, and mounting regulations to try to keep them away and control the situation. Otherwise, as whale-watching continues to expand, and as regulatory and enforcement agencies struggle to manage the mess, there are going to be many dissatisfied customers frustrated with the traffic levels and even jaded by any closeness or intimacy with cetaceans they do experience. This may be the fate of whale-watching—unless we change our way of thinking.

Close encounters have served a valuable role in getting people interested in cetaceans and their conservation, but are close encounters truly as compelling as seeing cetaceans interacting among themselves, undisturbed and undefined by human intervention? Is it more interesting to see a dolphin calf supported by its family and to glimpse complex social rituals—or to try to get as close as possible and then watch a dolphin or whale staring back at us? Yes, it is a thrill, and there is no proof that close encounters harm cetaceans, but we must accept that there is a limit to how many times and how much of their life they can spend entertaining humans—even if we cannot define the limit. We may be too conservative in our application of the precautionary principle—the idea that policy-makers in the absence of hard knowledge should take a cautious route. Still, this principle and all the regulations that might go with it mainly serve to erect fences between people and whales. Why not look positively and try to develop a novel approach, a new ethic when it comes to whale-watching?

My closing address at the conference on how best to watch whales and dolphins tried to do just this. First, I focused on the educational, scientific, and conservation values of whale- and dolphin-watching. I told a sea of eager Japanese faces that whale-watching was not as valuable in terms of education and science as many people say it is or imagine it to be. According to my 1997 operator survey, only 35 percent of the world's whale-watch operations have naturalists or nature guides on every trip. And only 9 percent of operators have researchers or naturalists who conduct research as part of their trips. Some 57 percent never conduct scientific research or offer information to scientists, but 44 percent worldwide would be willing to offer their boat free to researchers. This last statistic is encouraging, but it is depressing to discover that whale-watching is largely a commercial enterprise. The lack of education and poor value to science represents a massive wasted opportunity.

In my talk, I promoted the idea of "well-balanced whale-watching": for whale-watching to reach its potential and be successful, it needs to provide an enjoyable, educational experience that contributes to scientific knowledge and marine conservation, as well as being a profitable commercial venture. The most valuable whale-watching in the world, in terms of rate of return and socioeconomic value, is arguably around southern New England. The best operators here have made their trips into a thrilling scientific and educational experience: you get to see the scientists in action, doing their work from the boat, and you see the level of care they have toward the animals they study.

Of course, the definition of "education" varies by community, country, and culture. In Japan and Norway, where whaling still occurs, whale-watching offers another way to view cetaceans, if nothing else. Yet no matter the country or culture, learning more about cetaceans has the capacity to enhance the level of our appreciation and understanding of marine life and the natural world. Since my talk, I have continued to think about a new ethic for whale-watching. I think it is time to adapt an idea from land-based wildlife and wilderness tourism—that the best way to observe wild animals is to watch without being noticed, to become invisible, like the birdwatcher in the blind. It may not be fully achievable in the short term, but I believe it needs to be an idea that is recognized, articulated, established, and reinforced. With this new ethic, the goal of most whale- and dolphin-watching changes from wanting to get as close as possible and interact, to trying to watch, learn, and enjoy without disturbing. And a complementary aspect of this new ethic about visiting the ocean, adapted from land-based wilderness tourism, encourages us to tread softly, leaving the faintest "footprint" possible. In the sea, this means to watch from the cliff or shore when you can, to travel in boats with minimal wakes and unobtrusive sound profiles, to move quietly through the world, and to keep eyes, ears, minds, and sensitivities open.

I believe that, with this new ethic and with the goal of "well-balanced whale-watching" in the mind of every cetacean watcher, operator, and regulator, we can then truly celebrate the close encounters when they come. And they will come. Cetaceans will never ignore us, but the close encounter will return to its position of special, treasured experience. We might even sharpen our sense of wonder toward these extraordinary animals.

PART III

Beyond Our Myths:
The Real Life of Dolphins

In this section, we hear from people who have brought dolphins from legend to modern-day awareness in their efforts to protect them. In her poem "Hunger," internationally acclaimed Chickasaw writer Linda Hogan echoes the consumptive nature of our species toward other animals and the natural world.

Pat Weyer weaves together the mythic memory and symbolism of dolphins with her artist's experience in "Signs of Becoming." Weyer writes that "the relationship between human and dolphin is a primordial, deep, and abiding reality—not a lost fantasy."

In "Springer's Homecoming," Howard Garrett tells of an orphaned orca calf who was successfully reunited with her native pod, a moving drama that was international news. It also set a hopeful precedent for restoring other displaced orcas to their home waters.

In "A Sperm Whale: Reality Is More Magical Than Myth," Bill Rossiter shares a unique experience with a stranded whale who is eventually returned to the wild. Naomi A. Rose, a marine-mammal scientist and a powerful advocate for cetacean welfare, shares her personal exploration into the world of captive dolphins in "Sea Change." Rose has said, "Dolphins are not there for the same reasons we are—they are not there by choice." Another respected scientist, Lindy Weilgart, addresses what may be one of the greatest hazards to life in the ocean in "Acoustic Smog."

Betsy Smith's account of "The Discovery and Development of Dolphin-assisted Therapy" reflects her fascinating journey from working with captive dolphins to celebrating their wildness, as a way of healing both humans and dolphins. Biologist and nature writer Leigh Calvez shares what have been, for her, "Dolphin Lessons" in her insightful portrayal of spinner dolphin observations in Hawaii.

From a country where more swimmers go to encounter wild dolphins than any other place in the world, researchers Rochelle Constantine and Suzanne Yin document, in "Swimming with Dolphins in New Zealand," the long-term effects of such interspecies interactions. And finally, Marc Bekoff, a pioneering cognitive ethologist, adds his voice to those who passionately advocate dolphin protection in "Troubling Tursiops: Living in Harmony with Kindred Spirits."

From all these writers, we gain a sense of the ways in which our longing to be with dolphins in their own element might affect or change their lives.

Hunger

Hunger crosses oceans.
It loses its milk teeth.
It sits on the ship and cries.

Thin, afraid,
it fashioned hooks to catch
the passing songs of whales so large
the men grew small
as distant, shrinking lands.
They sat on the ship and cried.

Hunger was the fisherman
who said dolphins are like women,
we took them from the sea
and had our way
with them.

Hunger knows we have not yet reached
the black and raging depths of anything.

It is the old man
who comes in the night
to cast a line
and wait at the luminous shore.
He knows the sea is pregnant
with clear fish

and their shallow pools of eggs
and that the ocean has hidden
signs of its own hunger,
lost men and boats
and squid that flew
toward churning light.

Hunger lives in the town
whose walls are made of shells
white and shining in the moon,
where people live surrounded
by what they've eaten
to forget that hunger
sits on a ship and cries.

And it is a kind of hunger
that brings us to love,
to rocking currents of a secret wave
and the body that wants to live beyond itself
like the destitute men
who took the shining dolphins from the sea.
They were like women,
they said,
and had their way
with them,
wanting to be inside,
to drink
and be held in
the thin, clear milk of the gods.

Signs of Becoming

Log entry—July 3, 1988

I am lost in an aqua drift. The wooden hull creaks rhythmically against the strain of rope. Lighthearted chatter of the crew fills the decks of the rocking ship in a tranquil sea.

Suddenly a feeling in my throat percolates like soda water. I gaze out to see them approach. They pierce the flat perfection with one silent arc after another. I keep the news to myself. Looking up at the first mate I ask, "What happens when you find yourself living your dream?" He flashes a smile and replies, "You dream another dream." But this dream is not finished yet. I have been dreaming of dolphins all my life.

Thirteen years as volunteer artist for the Wild Dolphin Project in the Bahamas has been a spiraling dive into what was once only a recurring dream but has now become reality, a full embrace of the age-old fascination with another intelligent being.

How old is the relationship between man and dolphin? My search for the answer has been cyclical rather than linear, meandering through the patterns of mythic time. The image of a spiral has been the symbol of both my individual search as well as a powerful natural archetype. Spirals are primordial aquatic symbols, signs of becoming, used by ancient cultures as far back as the sixth millennium B.C. to denote growth, the life cycle, rebirth, and water.[1] The ancient Greeks made the connection between the dolphin and the life-giving powers of the sea. The evidence is embedded in early Greek language. Delphys, the Greek word for Dolphin, is related to Delphis, which means womb. "The Dolphin, then, is the living womb of the sea."[2]

Beginning my search, I found reference to the Sorokka Spiral, an ancient Finno-Ugrian petroglyph discovered on a cliff face near the White Sea in Russia. Traced along a counter-clockwise spiral, this petroglyph, dated 3500 B.C., depicts the life of a mythic hero figure. Traveling along the outer edge of the spiral, painted panels reveal what is most likely the hero's birth. Moving down along the spiral, tools for the hunt, small birds, mammals, and moose are displayed with images of other human figures and celestial symbols. The spiral ascends, perhaps signifying that hunting days are on the wane. Strange, enigmatic objects appear, and then the circular form turns inward again. At the central point of the spiral, the hero figure is portrayed at the end of his journey riding on the back of a whale/dolphin. This innermost point most likely signifies death and transformation. The ancient Finns appear to have used the innate generative structure of the spiral to express their totemic spirit and their belief in reincarnation. "Just as eating their prey was considered an act of taking the animal spirit into oneself, so reincarnation was likewise understood as a cyclical transformation from predator to prey back to predator, over and over again, forever."[3]

But these are not the oldest inscriptions of dolphins. Discovered in France, two dolphin figures incised on an antler from the Magdalenian culture dated 18,000–14,000 B.C. establish that Ice Age religion of northern Europe held both reindeer and dolphin as sacred animals involved in fecundity rituals.[4] Perhaps mythic memory of the relationship between man and dolphin spirals back even further. This great expanse of time is hard to fathom. How does the story of the mythic dolphin connect? Are the themes of transformation, reincarnation, fertility, and salvation regarding dolphins universal? These questions have framed and informed my art, my dreams, and my travels.

While studying in Florence, I found an old Renaissance-style print in a bookshop depicting the sea god, Poseidon, as a satyr with companion dolphins. Intrigued by these large-lipped dolphins with pompadour hairdos instead of fins, and noticing once again the recurrence of the dolphin motif throughout Mediterranean and Aegean art, I asked the book dealer about it. He explained that the ancient philosophers held the dolphin sacred and developed a body of literature called "delphinology," myths and legends about the relationship between humans and dolphins. Later Christian beliefs absorbed some aspects of delphinology, and in early Christian art the dolphin became a symbol for salvation.

Fascinated by my early discoveries, I began to seek out this literature on delphinology. My search led me from ancient scholars such as Plutarch and Pliny to the more modern works of Carl Jung and Joseph Campbell. I learned not only about the pervasiveness of myth-making in all human culture but also about how it functions in the human psyche. Campbell explains, in *The Way of the Animal Powers*, that the principal method of mythology is the poetic, that of analogy. For example, death as sleep; the light of the sun as consciousness; the darkness of caves, or of the ocean depth, as a womb or, alternatively, as death. There are many such analogies recognized the world over. The power of recurring patterns of images or dreams suggests a "collective unconscious" that binds together all conscious beings. Carl Jung, the famous psychoanalyst, referred to the collective unconscious as "all psychic contents that belong not to one individual but to . . . a society, a people, or to mankind in general."[5]

My own recurring dream, one that has been with me since adolescence, finds me engulfed in a deep void. I am swimming or flying in pure blackness. A strange sensation, a sizzling in my throat, alerts me to another presence with no discernible form. From this feeling emerges an image, and I am gliding past a huge old eye with great, sagging lids. I am not frightened but instead am filled with awe and continue to swim calmly in the abyss.

Years after this dream first began appearing, I would carve the images of the old eye and swimmer in a block of lead crystal. Now I understand that others may have shared similar dreams in the "collective unconscious." Uncanny similarities found in myths from the Maori, Amazonian, and Tlingit-Northwest Coast Indian cultures correspond in plot to the well-known Greek myth of Dionysos. All are transformation myths that involve the changing of cruel human beings into dolphins and whales that become friendly to man.

The Dionysos myth, mentioned previously in this volume by Ashley Montagu, falls under the category of etiological myth created by ancient peoples to account for natural or social phenomena. The theme of transformation is universal and identified as an archetypal construct used throughout delphinology to explain the story of the human-dolphin bond. The transformation theme also functions as an anthropomorphizing device to explain dolphin beneficence and intelligence. In "Legends of the Maori," Antony

Alpers identifies gentle sea creatures called "Taniwha" as dolphins. Similar to the scenario of the Dionysos myth, a murderer named Ruru is turned into a Taniwha, destined to live forever by the coast and to meet every canoe that passes by.

The theme of change surfaces once again in the Tlingit myth "The Origin of Killer Whales," recorded by John R. Swanton. A man named Natsilane constantly quarreled with his wife. His brothers-in-law undertook to protect their sister. They took Natsilane to an island far out to sea and left him. Natsilane carved whales from many different kinds of bark: cottonwood, alder, hemlock, red cedar. Every time, the wooden whales merely floated, unchanged from their wooden origins. Finally, he carved whales from yellow cedar, and these swam away as killer whales. The whales swam out a long distance and, when they returned, changed back into wood. He then made holes in each of their dorsal fins and, holding on to them, was taken out to sea. The next time he saw his brothers-in-law in their canoes, he put his whale spirits into the water. The whales smashed the canoes and killed the brothers-in-law. Natsilane then told the killer whales, "You are not to hurt humans anymore. You must be kind to them."

In Amazonian legend, an interesting reversal occurs in the theme of transformation. Candace Slater's book *Dance of the Dolphin* documents narratives that involve the metamorphosis of freshwater dolphins into human beings. Unlike many other traditions, dolphins *(encantados)* exhibit a darker side as enchanted beings and tend to abscond with the object of their desire to an underwater city *(Encante)*. "Dolphins turn into people and show up to dance at parties. My own grandmother once danced with a dolphin, so I know for sure that these things happen. At least they used to happen. Today, it is more difficult."[6] Slater explains in her footnotes that a small number of human-to-animal metamorphoses occur in narrative form.

These narratives follow the structure of the Dionysos myth. For example, a story from Ecuador describes an unfaithful wife and her lover who turn themselves into dolphins after the woman's husband discovers her infidelity. And a Colombian story recounts the change of badly behaved villagers into dolphins in the wake of a flood that sweeps them away during a great festival.

But it is the Australian Aborigines who embrace the full flowering of the archetypal theme of transformation. Aborigines regard themselves as the direct descendants of dolphins. The natives of Groote Eylandt in the Gulf of Carpentaria celebrate their mythic heritage by chanting and dancing until they enter a trance-like state called "the Dreamtime," a place where humans commune with dolphins to reaffirm their ancient bond. Almost two thousand years ago, in an essay titled "On the Cleverness of Animals," Plutarch wrote about the human-dolphin bond: "To the dolphin alone, beyond all others, nature has given what the best philosophers seek: friendship for no advantage. Though it has no need of man, yet it is the friend to all men and has often given them great aid."[7]

Log entry—July 5, 1988
In a cloud of aquamarine, I struggle to clear my mask. I hear the clicks and whistles of the dolphins, but I only sense their presence. I cannot see them. Finally my mask clears. My breathing calms. Turning to the side, I look directly into the red sagging lids of an old eye. A dolphin returns my gaze, and we are locked in mutual admiration. I feel strangely comfortable, not at all frightened. The spotted dolphin and I swim a slow dance that feels like an old memory.

As I surfaced, images of the crystal carving and the recurring dream danced in my head. Back on the boat, I am told that I have encountered "Romeo," the elder male of his family group. For thirteen years, Romeo and I continue to meet. I can sense his presence before I even see him. I feel an inexplicable bond with this dolphin.

Log entry—June 25, 1992
Everything quickens. My heart is racing and my head is buzzing as we make the jump. Large swells tug at my fins. Once I hit the water, a fast joyful feeling envelops me. In happy reunion, the dolphins dart excitedly between the other swimmers. Always hoping to encounter Romeo, I think I see him among the larger animals. But I am not sure it is him. Distracted, I swim alone. Something brushes my shoulder—just a piece of seaweed. Then Romeo slips close by my side, as if to say, "Wake up! Of course it's me." He turns and comes head to head, then drops beneath me hanging in a vertical position. I am suspended above him with my arms outstretched.

During a recent trip to Greece, I had the opportunity to visit a dedicated team of dolphin researchers working in the Gulf of Corinth near the ancient temple of Delphi. This trip offered new insight into the historical record imbedded in theogony myth (myth relating to the birth and genealogy of the gods) within the fabric of delphinology. On this routine scouting trip to photo-document dolphins in the Gulf, an intriguing intersection of mythic and real time emerged, and left me wondering—could these dolphins be the descendants of the mythic dolphins of Delphi?

Log entry—September 10, 2001

It is midday in the Gulf of Corinth. I dreamed of coming here, but gave no thought to the relentless pounding of the Greek sun. Finally they come streaming through eons of sun-drenched turquoise sea. Across the strait looms Mt. Parnassus magnified in pulsing ochre heat. Striped and common dolphins play together on the bow. As the dolphins dive, their elegant forms bend in refracted light, and their characteristic stripes become yellow squiggles at depth. I realize—of course! This same yellow squiggle on the dolphin's side is depicted in the dolphin frescoes at the Palace of Knossos. The dolphins continue to dive deeply into my thoughts. I glimpse this scene with an ancient eye. Surely, the famous dolphin frescoes at Knossos were reflections of true observation, testaments to Minoan artistry, and gave witness to the dolphin's presence in these waters. I breathe and take in this fabled place where myths were born, and where Apollo the Sun God defeated Delphyne the dolphin/womb monster to become Delphinios ("dolphin-god") of Delphi (Dolphin Town).

In *Mind in the Waters: A Book to Celebrate the Consciousness of Whales and Dolphins*, assembled by Joan McIntyre, Charles Doria comments, "An encounter between a strong god or hero and a snake/dolphin/woman crops up in so many places I am beginning to suspect this must have been an integral part of a once-universal Mediterranean creation story."[8] The writings of classical scholars such as the historians Plutarch and Herodotus, the poets Pindar and Oppian, and the naturalists Aristotle and Theophrastus provide detailed accounts of dolphins rescuing men from drowning, piloting ships to safe harbor as well as befriending young boys, and helping fishermen at work.[9] These historical encounters, combined with the primordial symbols of delphinology, have deeply etched the image of the dolphin into the human psyche.

Log entry—July 9, 1988

I swim in the slipstream of a joyful pair of juvenile dolphins. A sound jogs me to the surface. The dolphins shudder then regroup at my side, turning me back out to sea. I crane my neck to recognize the boat, a tiny tossed water toy over one mile away. Its anxious bell beckons me back. The dolphins press forward, and I want to go with them because I can see at least forty more dolphins ahead. The young spotteds are leading me toward their family group, but I am torn between their homecoming and mine. I am not afraid to swim the distance but feel alarmed by the insistent signal bell to return to the boat. Now I hesitate. The water churns and one dolphin sidles up close and looks me in the eye. I have to go home now—I thought to myself. In that instant, the dolphins dive beneath me, one under the other, presenting their dorsal fins. Though I do not hold on to their fins, I am carried by the dolphins' pressure wave. They return me safely to the boat. I am a dolphin-rider! Later, the crew reports that all other swimmers escorted by dolphins were returned to the boat at the same moment.

The relationship between dolphin and rider is complex. Concrete evidence of actual dolphin-riding is documented in both writing and statuary at sites throughout the ancient Mediterranean.[10] And there is the mythological connotation of dolphin-riding as safe passage from birth to rebirth.

Log entry—September 10, 2001
    After the dolphins depart, our serene passage through the Gulf of Corinth turns into a wild carnival ride. The water turns to steel and begins to boil in the wind. As we careen through six-foot swells, I feel the full embrace of my new Greek friends. Together we brace for the storm. I struggle to hold on to this most perfect day with the dolphin researchers, but the weather churns the narrow strait into spin-drift and spume. Our helmsman plows through each wave, determined to reach the harbor before nightfall, but we come to a silent dead stop. The boat gyrates as we frantically check fuel lines. Finally, the engine starts and labors on toward the port. Everyone is quiet, cold, and absorbed in private thoughts of home. The water, now black and large, is staring me in the face. In my blurry head, I conjure up the mythic tale of Arion, who, in ages past, was forced to jump into these same tormented waters, forsaking all hope until dolphins, attracted by the sound of his harp, rescued him and carried him safely to the Cliffs of Taenarum. Just the thought of these mythic dolphins comforts me. The sky begins to clear as harbor lights play on flat water. We make it to shore and tumble off the boat.

This fifth-century recreation of the Hymn to Poseidon celebrates the rescue of this most famous dolphin-rider, the poet Arion. (Arion's original poem no longer exists.)

Arion: Hymn to Poseidon

Sea lord Poseidon
golden trident
biding earth in the child-swollen salt sea
you most high
whom fish encircle
dance about lightly
fins up down
back forward
snub nose manes rippling
running hard, sea pups
dolphins music lovers
briny kids
the girl Goddesses
Amphitrite's
Nereids
Milk breast-fed
Whose hump backs I rode
To Taenarum Cliffs
In Pelop's land
Furrowing the flat sea plains
A trackless way
The time the trickers
Threw me off their smooth ship
Into the swellings of salt purple sea.

      (in McIntyre, ed., *Mind in the Waters*, 37)

Images of Aphrodite as dolphin-rider abound. Aphrodite inherited her dolphin-riding status from the ancient lineage of the mother-goddess associated with the primordial emblems of fish, for fertility, and lozenge, for womb. Neolithic sculptures dated 6000 B.C. from northern Yugoslavia link the images of fish and lozenge that would later become attributes of the goddess.

In Sanskrit, the term for the goddess of love translates to "she who has a fish as her emblem." Buffie Johnson explains, in her book *Lady of the Beasts*, that "the dolphin's womb, wisdom and compassion for humans connect it to the Great Mother; later, the relationship is transferred to Aphrodite."[11] One particularly vivid account of Aphrodite's birth involves an egg fallen from heaven into the water saved by fishes, who bring it to shore, where it is hatched out by a dove. The link between the goddess Aphrodite, the dolphin, and the symbol of the egg is illumined by discoveries made in Mycenean shaft graves from the second millennium B.C. containing ostrich eggs with faience dolphin appliques. Eunice Stebbins, in her book *The Dolphin in the Art and Literature of Greece and Rome*, explains the significance of this find. "The egg is a symbol of future life, its surface representing the water over which the dolphins convey the dead, and at the same time the symbol of the universe from which springs the goddess who is carried ashore by the dolphin after her birth from the sea-foam."[12] These concepts, used in both pagan and Christian religions, are reiterated in the *vesica piscis* ("vessel of the fish") reflecting the idea of the feminine as vessel.[13] It is important to note that ritual vessels with dolphin iconography have been found in Minoan tombs. These vessels, held sacred by their creators, evoke the *vesica*–dolphin/sea–womb symbolism and project complex ideas on the origin and nature of life.

Aphrodite makes her debut as the Fish Goddess Atargatis in the Syrian temple at Ascalon, her most ancient shrine. She is great ancestor to the mermaid image of today. In pre-Hellenistic thought, Aphrodite was identified with the moon and the dolphin, and she held the role of guiding souls to the underworld. In ancient Greece, coins stamped with dolphin images were placed in the palms of the dead to ensure them safe passage through the afterlife. Johnson continues to explain: "In addition to having specific associations with the womb, the fish represents the mother-goddess herself and often appears as her mount. The animal incarnation on which the transformed deity rides is not merely the vehicle but actually an alternate form, an epiphany of the divinity. The earliest images of the deity were often in animal form and only later assumed anthropomorphic shape."[14] As the mythology of the dolphin-rider unfolds, we see the human form elaborated out of the animal and then at times imposed as a rider on the original beast.[15]

Even though Aphrodite enjoyed a long reign as dolphin-rider, Eros, the male god in the persona of her son, would appropriate her place as the dolphin-rider in Greek mythology. This transfer of worship from goddess to god appears approximately in the early seventh century B.C. and accounts for the proliferation of dolphin images associated with the phallus, sun, sky, and wind. These male aspects are personified in Apollo the dolphin god, and later on in Greek mythology as Triton, son of Poseidon and Amphitrite. Depicted as half man/half dolphin, holding an erect dolphin in his hand, Triton was the phallic dolphin, "wide-ranging and violent,"[16] who assumed the female powers of dolphin and sea snake. Like Aphrodite, Triton was associated with the wind. The riders may have changed but the dolphin continues to spiral through time, symbolizing the archetypal themes of transformation, reincarnation, fertility, and salvation.

The most astonishing manifestation of the dolphin-goddess is documented in an obscure book, *Deities and Dolphins*, by Nelson Glueck. High on a desert hilltop in TransJordan, the ruins of a Nabataean temple at Khirbet Tannur reveal a stone goddess with a tiara of dolphins on her head. One well might wonder what dolphins are doing in the desert. Glueck explains: "For much of the first millennium B.C. and well into the first millennium A.D., representations of deities with dolphin symbols or of dolphins by themselves are found from Iran to England in sculptures of stone or bronze or pottery or on coins, lamps, mosaics, and murals. Much religious importance was attached to them for safety and succor on the journey through life first by sea and then by land and perhaps even greater significance for assistance on the much more perilous, uncharted routes through the world of after-life."[17] Glueck also points out two major factors that contributed to the Nabataean dolphin cult. First, the importance of the dolphin was heightened by the astral significance appointed to it in the ever popular zodiac. Second, the nomadic life of the Nabataean spice traders brought them to seaports throughout the Mediterranean, exposing them to dolphin images venerated in Hellenistic Europe, Africa, and Asia.

Perhaps most powerful of all these images is the ancient Minoan expression of the dolphin, seen as the integral whole, swimming alone in its natural environment. The best example of this is a large mural in the Palace of Knossos on the island of Crete. Running spirals create a buoyant border around a painted mural, depicting dolphins playing and weaving among sea plants and fish. Curiously, these dolphins, though stylized, are rendered in a far more naturalistic way than the large-lipped dolphins of the Renaissance. Recent archaeological excavation of wall paintings and portable artworks (ritual vessels) at Akrotiri on the Greek island of Thera reveal extensive use of dolphin iconography and contribute major evidence of the dolphin's exalted status in ancient Aegean culture. These frescoes also document the Minoan inclination to render nature as faithfully as possible. Christos Doumas, in *The Wall-Paintings of Thera*, states, "This predilection for realism seems to be associated with the narrative character of virtually all the compositions."[18] It seems the Minoan artists from the marine culture of the Aegean were inspired to document and celebrate their encounters with nature. They were far more intimate, knowledgeable, and comfortable with their animate world than the later Renaissance artists. These later artists were estranged from nature through experience of the Dark Ages and by the impact of Christianity, which demoted the importance of the natural world within the spiritual hierarchy. Renaissance images of dolphins tend to be fantastic and sometimes even monstrous. This change in perception of the natural world also correlates with the absence of anecdotal accounts of dolphin encounters in Western literature from the second century A.D. lasting well into modern times.

Log entry—July 28, 1998
   We drift all night over the edge of the bank into water that reaches depths of 850 feet. It is my first night dive while the dolphins fish for squid, drawn to the bright lights placed on deck. At midnight, the dolphins streak around us in pursuit of squid and needlefish. Three of us get into the water. It feels like a great seafood soup. At first the dive is disorienting and scary, but the energy of the dolphins overrides my anxiety. I find myself eye to eye with ecstatic dolphins cartwheeling underwater, then up into the air. We share the thrill of the chase. Special ear phones used by professional divers help us to hear the dolphins' high-frequency emissions. The sonic bursts, click trains, and whistles of the dolphins surround and penetrate my body. The dolphins ascend and descend in the water column with incredible speed. They seem to ride the beams of light from the boat like an express elevator. This is the wildest encounter of my life.

   Mythology provides a glimpse into a time when a greater intimacy existed between human and nonhuman animals. Looking back through the spiral of mythology, there is compelling evidence of a dream lived. Interspecies interaction is not only an ancient legacy but also an ongoing relationship, one that extends the boundaries of self and consciousness. The relationship between human and dolphin is a primordial, deep, and abiding reality—not a lost fantasy. But I wonder whether, while we struggle to recover and recreate this dream, the dolphins may simply—and easily—remember.

NOTES

1. Buffie Johnson, *Lady of the Beasts* (San Francisco: Harper and Row, 1988), 234. Also see Marija Gimbutas, *The Language of the Goddess* (San Francisco: Harper and Row, 1989), xix, xxii.

2. Joan McIntyre, ed., *Mind in the Waters: A Book to Celebrate the Consciousness of Whales and Dolphins* (New York and San Francisco: Scribners and Sierra Club Books, 1974), 33.

3. See Whalewatching Web (www.physics.helsinki.fi/whale/comics/nollman/beluga.html) and "Finno Petroglyphs and Beluga Whales," an excerpt from Jim Nollman, *Not Talking to Beluga* (New York: Henry Holt, 1998).

4. Johnson, *Lady of the Beasts*, 228.

5. An excerpt from Daryl Sharp, *Jung Lexicon: A Primer of Terms and Concepts* (Toronto, Canada: Inner City Books, 1998), introduction.

6. Candace Slater, *Dance of the Dolphin: Transformation and Disenchantment in the Amazonian Imagination* (Chicago: University of Chicago Press, 1994), 59.

7. Jacqueline Nayman, *Whales, Dolphins and Man* (London: Hamlyn, 1973), 41. For a similar quotation from Plutarch's philosophical work "Moralia," see also Amanda Cochrane and Karena Callen, *Dolphins and Their Power to Heal* (Rochester, Vt.: Healing Arts Press, 1992), 57.

8. Charles Doria, "The Dolphin Rider," in McIntyre, ed., *Mind in the Waters*, 35.

9. See Ashley Montagu, "The History of the Dolphin" (in this volume).

10. See Brunilde Sismondo Ridgeway, "Dolphins and Dolphin-Riders," in McIntyre, ed., *Mind in the Waters*, 38.

11. Johnson, *Lady of the Beasts*, 239.

12. Eunice Stebbins, *The Dolphin in the Art and Literature of Greece and Rome* (Menasha, Wisc.: George Banta, 1929), cited in McIntyre, ed., *Mind in the Waters*, 43.

13. Johnson, *Lady of the Beasts*, 224.

14. Ibid., 244.

15. See Doria, "The Dolphin Rider," who states: "In other stories the woman and the snake/fish/dolphin are separate creatures, or rather the human form is elaborated out of and then at times imposed as a 'rider' on the original beast" (35).

16. See ibid., 43. Doria quotes Hesiod's description of Triton as *eurybias*, "wide-ranging and violent."

17. Nelson Glueck, *Deities and Dolphins* (New York: Farrar, Straus and Giroux, 1965), 413.

18. See Christos Doumas, *The Wall-Paintings of Thera* (Athens, Greece: The Thera Foundation Petros M. Nomikos, 1994), 27. Doumas describes Aegean wall paintings as "little essays on nature" and explains that "the realistic portrayal of the elements of the flora and fauna of these regions bears witness to direct contacts and the actual presence of Therans there."

Springer's Homecoming

When an orca calf was found swimming alone in Puget Sound, Washington, in the winter of 2002, no one knew what to do. Where was her family pod? Why was she left behind? Was her mother dead? When human orphans are left behind on a doorstep, there are procedures for providing care and placement in foster families so the youngster will learn the ways of our culture and have a chance to grow and become a member of society. But this was another species, and there were no institutions to take care of her, no idea at all of how to help prepare her for membership in her orca society.

In the Pacific Northwest, we have learned to care about the orcas that swim, forage, and play in our urbanized waterways. Over the past few decades, we have grown fond of J, K, and L pods of the Southern resident orca community that appear in lower Puget Sound in winter months and around the San Juan Islands in spring and summer. We know they have indelible family bonds and that they call out to one another constantly, using vocal traditions unique to each community. And we know they have never hurt a human in the wild despite decades of being shot at and trapped for delivery to amusement parks. Throughout the Pacific Northwest, we share a sense of respect and affection for these magnificent whales, and we feel we owe them something for our thoughtlessness in years past.

But who among us knows anything about raising an orphaned orca? First, some fact finding was needed to try to locate this little whale's parents or, in this case, her pod. There were two immediate possibilities. The first guess was a result of the timing of her appearance, just two weeks after an adult female "transient" orca had died in a stranding outside Puget Sound, at Dungeness Spit. Though residents and transients look very much alike and are considered to be the same species, they are members of entirely different orca societies. Transients have specialized in eating marine mammals such as seals, sea lions, porpoises, dolphins, and even large whales, whereas residents eat fish. Transients keep their pods very small, usually just three individuals, whereas residents travel in much larger groups. All three Southern resident pods often join together in festive gatherings. Residents are more nonchalant or even gregarious around people, whereas transients are more vigilant and evasive.

A young adult male orca at the scene of the stranding, probably the son of the dead female, beached himself repeatedly, but after three days rescuers were able to get him moving in the Strait of Juan de Fuca. Some reports told of a small orca swimming alongside the female just a day or two before she died, so the public had been alerted to watch for an orphaned baby orca. In fact, about the time the female was discovered, there were reports of a very small orca in, of all places, Swinomish Channel, near La Conner. In early January, a tugboat captain in Puget Sound reported a small orca to the Orca Network sightings network.

The day after the baby orca was found near Vashon, Ken Balcomb, who has been personally surveying the Southern resident orca community for over a quarter-century as director of the Center for Whale Research, went out in a small skiff to take a look at the calf. Balcomb declared that she acted much more like a resident than like a transient.

Balcomb had yet another clue that this was a resident and not a transient. To avoid an anticipated public frenzy, he and a half dozen veteran orca researchers had kept a secret about an astounding and unprecedented event that was unfolding in the remote reaches of Nootka Sound, on the west side of Vancouver Island. A two-year-old calf from the Southern community, known as L98, or "Luna," had somehow become separated from his pod and had stationed himself in a small, well-defined area of about two square miles for at least the previous six months. Luna had been written off as dead, because in almost thirty years of field research, no orca had ever left its family and been seen again.

Balcomb at first assumed that Luna must have repositioned himself in Puget Sound at the Vashon ferry dock, back in his family's home and habitat. But within a few days came the report that Luna was still 400 miles northward and had not budged an inch from his apparently self-imposed, imaginary enclosure. So the Vashon calf must be somebody else's.

Because the little orphan's saddle patch, an area just behind the dorsal fin that is used to identify individuals, had not yet fully formed and was obscured by a discoloration typical among orca infants, photographs were not useful for tracking down the baby's family of origin. So it took a bit of whale magic to reveal the solution to the mystery of the Vashon calf.

We know that each orca community uses its own distinct vocabulary of calls, much like language, ranging from melodic whistles to raspy buzz sounds, and we know that each pod within each community uses a few of its own calls, perhaps as a quick and easy identifier for recognition from miles away. So researchers dropped a hydrophone into the water near the calf and eventually recorded a series of fairly distinct calls.

The recording was sent to Helena Symonds at OrcaLab, where the Northern resident orca community had also been studied for over twenty-five years. Symonds spent hours listening through decades of recorded orca sounds to find one whale call recorded in 1988. When she found it, she knew right away who the calf was and who its mother was. On one tape was the calf—known as A73, or "Springer." On the other tape was her mother, recorded fourteen years earlier. Played side by side, even an untrained ear can hear the similarity. Acoustic traditions are passed on from mother to offspring, and it was a perfect match.

Now at least we had some history. Springer's great-grandmother, A10, had two daughters. One daughter, A24, was Springer's grandmother. A10 and at least one other whale in Springer's pod were shot that summer by persons unknown. Great-grandma died over the winter after the shooting.

A24 had seven offspring, but only two are still alive. High levels of PCBs and DDTs may have affected her ability to reproduce. A45, Springer's mother, was born in 1983. A45, like her mother, lost her first calf shortly after birth. Springer, A45's second calf, was born prior to the summer of 2000. They were seen together many times that summer in Johnstone Strait. She and her mother were last seen together in September 2000, near Ketchikan, Alaska.

In midsummer of 2001, A4 pod returned to Johnstone Strait without Springer or her mother. Both were feared dead. However, in late summer Springer was seen traveling with an adolescent female in one of the G pods. Springer was known to travel with G pod for at least a month.

At some point between the fall of 2001 and January 2002, Springer became separated from G pod and wandered into Puget Sound. Orca pods travel day and night and may cover 100 miles each twenty-four hours. Young calves keep up with the pod by riding alongside their mothers' flanks, where the waves are parted and resistance is reduced. Without that help, they would not have the strength and endurance to keep up. Did the adolescent female in G pod have a calf of her own and have to leave Springer behind? Did G pod at some point have to make a long run for more productive waters in Alaska, leaving Springer in their wakes? Was Springer unable to catch enough fish to provide sufficient energy to keep up the pace?

We do not know the answer, but we do know that by mid-January Springer had arrived at a spot near the Vashon ferry dock, at one of the best fishing holes for steelhead salmon in Puget Sound. No scientist would have speculated that an orca calf so young could have caught fish and taken care of herself, but Springer proved she was up to the challenge, as had Luna up in Nootka Sound.

Springer also found companionship. Bonds with their maternal groups for both males and females among orcas last a lifetime, unlike any other mammal known. In captivity, orcas captured from all over the world create social patterns and build trusting relationships among themselves and even with their human trainers. In addition to abundant fish to eat, Springer had ferries full of fascinated commuters moving by at regular intervals. Distressed media reports told of Springer's unnatural attachment to the *Evergreen State*. Mark Sears, a veteran researcher and compassionate orca observer for decades, lived next to the West Seattle ferry dock and was able to visit Springer almost every day over the winter. She had found a pretty good spot to camp out.

Now enter the rescuers. Springer is a marine mammal and a member of an orca society, albeit separated from them. We humans are land mammals and members of our modern industrial society. Many of us, however, believed we knew what was wrong with Springer, and an array of dangers and maladies were discussed that would require us to intervene on her behalf. Veterinarians from marine parks assessed the situation and quickly diagnosed several potentially life-threatening illnesses. It was assumed Springer was not eating well, and when someone got a whiff of her breath and smelled a paint thinner–like aroma, it followed that she was burning up the energy stored in her fat supplies, which meant she was starving.

At least three marine parks offered to catch Springer and take her to their tanks, where they would presumably fatten her up. Whether they would ever release her back to the ocean was left a bit vague. She would be a spectacular money-maker, but that aspect was not part of the proposals.

Then observers began to notice that Springer was catching fish, and plenty of them. She often raised her head vertically above the surface in a spy-hop, holding the bloody carcass of a two-foot steelhead for all to see. She was not starving after all.

At first, Springer's discolored skin was described as normal for a young calf, but as the months wore on, several observers cried loudly on television that there was danger it would become seriously infected and possibly cause her death. From the start Springer was playing with driftwood logs like bathtub toys. Most veterinarians at first described this behavior as a result of missing her mother. Orcas are very physically engaged with one another. They often travel just inches apart, apparently in tactile contact as they swim. Young calves are especially prone to wallowing all over their mothers' bodies. Springer was without kinfolk or any warm-blooded companions, so rubbing along driftwood logs may have helped fulfill that need for closeness. Soon, however, the rubbing was said to be due to irritation from "whale pox."

She also came willingly to boats to eyeball the passengers and sometimes engage in little games. When Ken Balcomb first encountered Springer in mid-January, he frivolously made a hand motion signaling Springer to roll over, which she did perfectly at the first suggestion. She even held herself upside down, to Balcomb's appreciative astonishment. Springer, like other solitary dolphins, seemed eager to interact with her human visitors.

The National Marine Fisheries Service (NMFS), ultimately responsible for deciding what to do about Springer, held two public forums to hear opinions and sense the pulse of the community. The first and clearest message was that people would not tolerate Springer being taken to any concrete tank. The NMFS accordingly turned down offers from marine parks that wanted to take her.

In late April and early May, in a series of conference calls, a scientific panel advising federal fisheries officials said they needed to get a blood sample from Springer. Most of the scientists said she would probably need medical treatment and should be captured. In the absence of real symptoms, the NMFS said it had no such intention. On May 8, Springer obligingly waited alongside a small boat while scientists drew blood from her back. The blood tests showed no signs of disease or genetic defects, another disturbing prospect raised by the veterinarians.

In April and May, the calls to save Springer rose in volume. Her skin ailment, though common among killer whales, was "worsening." She was described in the media as "distressed," "ailing," "languishing," and "stranded in Puget Sound," although she was eating well, actively moving around, spyhopping, and breaching, and she had arrived of her own volition and was free to leave. Washington State Ferries officials became concerned that they might accidentally run over Springer, although from the start she was alert and agile. Roundworm parasites were seen in her feces, leading to anxiety about another death-dealing sickness that must be remedied by human intervention, although worms are almost universally found in marine mammals. Her attraction to boats became defined as dangerous and addictive on the assumption that she would prefer boats to orcas when offered the choice. One alarm seemed to trigger the next.

Several nonprofit groups asked the NMFS to capture Springer and transport her to northern Vancouver Island, where her pod was expected to return in early July. Though some scientists quietly disagreed, the groups persisted. A floating net pen was promoted as a way to tow her the 400 miles to her family. Headlines such as "Scientists: Sick Orca Needs Help" (*Vancouver Sun*, May 13, 2002) fueled rescue efforts. In May the Vancouver Aquarium pressured the NMFS to act soon on a capture plan, giving the agency until May 15 or it would withdraw an offer to help. Aquarium president John Nightingale warned that, unless the whale was captured and her medical problems remedied, "It's going to die, day by day, on the evening news." The chorus of outspoken opinions almost uniformly concluded that Springer was dangerously ill and needed to be captured for evaluation and treatment of these supposed diseases. Springer needed to be reunited with her family, but with that rationale for the capture already in place, any marine park veterinarian's opinion built on that belief would sound believable. It looked like Springer might end up in a tank after all, for treatment of medical problems she did not actually have.

On May 25, the NMFS announced that Springer would be captured, evaluated, and treated for at least two weeks. If the fisheries service and Canadian authorities felt the calf could be released without harming her native pod, she would then be transported to Johnstone Strait, roughly halfway up the northeast side of Vancouver Island.

On June 13, Springer was lured to a capture boat. The previous day they had led her into a floating pen by scratching her back and sides and offering to play with her with driftwood toys. On the capture day we watched from a ferry as Springer seemed to tease her captors, remaining just out of reach. She breached or half-breached six times in the hour before her capture. The seas were too choppy to use the floating pen, so they reverted to the soft lasso around her tail. She bolted and thrashed for a few seconds when the noose was tightened, but divers jumped in and held her tight to calm her down. Within minutes she was in a sling and being hoisted aboard a barge.

After having been seduced with loving touches, immediately after her capture Springer's blood was drawn, swabs were poked down her blowhole, and her urine was sampled. She was taken to a NMFS lab a few miles away, where an Atlantic salmon fish farm offered a pen and a supply of live fish, delivered down a long tube to avoid her associating the fish with humans. It is possible she assumed a connection between her captors and the fish anyway.

According to conventional biological thinking, it is necessary to minimize human contact with an animal that is proposed for return to the wild, lest it become habituated to human contact and thereafter refuse to rejoin its own kind. Thus the capture team distanced themselves from her after the capture. They stayed out of the water and observed her from a duck blind beside her pen. Voices were kept to a whisper to avoid being detected by the impressionable youngster.

Immediately after the capture, in a NMFS press conference, Dr. Jim McBain, head veterinarian for Sea World, emphasized the seriousness of Springer's health problems. "Her condition is a concern. This is not a robust killer whale." McBain also doubted whether she could ever rejoin her family. "To me, this is a big question now—is she going to know she's a killer whale and go with those animals?"

Springer's family, the A4 pod, was likely to arrive in mid-July and stay for only about two months. David Bain, orca researcher and former biologist for a marine park, said: "Her health is so poor right now there's no way she will be ready that quickly." Bain said it could take as long as two months of treatment before the whale was well enough for release.

When we saw these statements from recognized experts whose judgment would determine Springer's fate, we at Orca Network and other groups sent a message to the NMFS and the involved nonprofit groups saying that reports of Springer's medical problems were greatly exaggerated and could lead to her permanent captivity. Within a week the story changed completely. A new vet, Dr. Pete Schroeder, now spoke to the media, saying the acetone smell on her breath had nearly disappeared. The problem with worms also abated, and on June 21 federal fisheries officials said she looked healthy enough to be shipped back to her native Canada. In the meantime, her caregivers installed bathtub toys, such as her favorite stick and a driftwood log, to stave off depression.

Soon the orca's handlers were saying again what they said in the first few days after Springer was discovered, that the skin irritation was probably due to the stress of loneliness, noting that a reunion with her long-lost family could cure it. "She's responsive, she's bright, she's alert, and she's sensitive to a lot of things in her environment," said an upbeat Schroeder. "She has pretty close to a clean bill of health now. Her best move now would be to take a trip north." Two more tests also showed that the orca did not have any inborn genetic defect affecting her metabolic system, as the NMFS scientists originally feared.

On July 2, NMFS regional administrator Bob Lohn announced that Springer had passed all medical tests and was ready for transport to Canada for release with her family. Lohn explained how orcas act when meeting up after a separation: "When killer whales enter the strait, there's a sense of joyous reunion as they squeak and call out, often leaping almost completely out of the water." Springer would finally be among those who really understood her.

On July 13, in an unprecedented and widely televised journey, the little calf was lifted aboard a high-speed catamaran donated by a Whidbey Island boat builder and piloted by a local whale-watch captain. With regional and national news agencies looking on, she was carried in a container half-filled with water and and ice for the 400-mile, twelve-hour journey to her home waters off Telegraph Cove, British Columbia. There Paul Spong and Helena Symonds of OrcaLab announced that, just days before Springer's arrival, her own family pod had returned from Alaska to their summer feeding grounds off Hanson Island. Springer's grandmother (A24) and great aunt (A11) were there, as if awaiting the return of their orphaned calf. Everyone anxiously awaited the reunion.

Springer's arrival at a temporary floating pen in a bay on Hanson Island was greeted by a pair of Namgis Indian canoes, including one with a banner reading, "Welcome Baby Orca." The Namgis had been given a special permit to catch seventy-five live salmon for release into the pen for Springer's dinner and homecoming gift. As soon as she was put in the pen, Springer immediately vocalized. "She spent a lot of time doing very energetic breaches. She did every kind of breach you can imagine. She did breaches on her side, jumped on the water and landed on her back, did belly flops, a lot of spyhopping, looking around. It looked for all the world like she was absolutely thrilled to be home," said David Huff, Vancouver Aquarium veterinarian. Huff said Springer spent the night hunting the live salmon.

At 1:30 A.M., scientists at the pen got a call from OrcaLab indicating that a pod, including Springer's immediate family, was vocalizing while traveling up the passage in the direction of Springer's pen. About the time the call came in, Springer became very excited and began to breach and call very loudly while nudging the net in the direction of the calls. Scientists reported that the calf "practically blew our headphones off." Springer and the pod seemed to interact vocally. Springer stayed very excited all night after the encounter.

Early the next afternoon, the whales returned to the bay. Immediately Springer made an "obvious vocal connection" with the A35 pod, led by A11, Springer's grandmother's sister. They milled and mingled for a few minutes, then lined up side by side facing Springer. They gradually approached within a few hundred yards of her net pen. Springer again showed that she wanted to go by vocalizing and pressing on the net, and the team agreed that the time was right. Scientists pressed three small suction cup radio tags to her to enable them to track her movements for a few days. Springer grabbed a fish in her mouth but did not eat it, as six divers held her tight. The net was dropped and they guided her out of the pen into the open water to join the pod. There was not a dry eye around the pen as Springer swam off, holding the salmon in her mouth, seemingly as an offering or a sign of trust. From this point on we humans were no longer players but merely observers as the unfolding drama involved only Springer and her family. She went charging off toward the pod, but then, about a hundred yards away from them, she stopped and played with some kelp. Apparently rejoining your orca pod after a long absence is not automatic, even when you know the right calls. We have no way to know the nature of the interaction between Springer and A11's family, but the pod moved slowly to the east while Springer turned gradually westward.

Researchers kept their distance to avoid disturbing whatever was taking place between Springer and her family, but on a few occasions Springer was seen following a few hundred meters behind other whales from her pod. Would they let her join them, or was she destined to remain a loner for the rest of her days? Or did she just lack the stamina to swim at speed with them? Everyone breathed a little easier on July 18, four days after her release, when she was seen sandwiched between a sixteen-year-old orphaned female named A51 and her younger brother, A61, like they were her bodyguards. She stayed very close to them for over a week. Twice A51, like a strict but loving parent, actively intervened as Springer headed off toward boats. Apparently the rule is: real whales don't play with boats. Then they all disappeared for over a month until August 25, when Springer was seen in the company of her grandmother's sister and other members of the A4 pod. "She's basically touring around with the A4 pod," said Symonds. Springer's reunification with her family seems to be a complete success. Springer's fans worldwide can follow Springer's travels with the A4s from written reports, live video, and even audio via www.orca-live.net. Unlike her human friends, she and her orca family share cultural traditions and understand each other's calls. She is finally back among those who really know her.

For six months Springer kindled our concern and compassion for her welfare. Endless conversations, conference calls, meetings, media reports, and talk shows followed her every move. At every turn of her story, interpretations of her behavior, health, and future prospects spanned a wide spectrum. In the end, a diverse assortment of sometimes differing viewpoints and agendas produced one collective action: Springer was given a ride home and offered a joyful reunion with her family. The moral of the story is a beautiful show of respect for orca freedom and family life. It is an extraordinary case of humans observing and honoring the way of life of an orca and her community.

BILL ROSSITER

A Sperm Whale: Reality Is More Magical Than Myth

Perhaps the sperm whale is really a genius in disguise;
the possibility cannot be totally discounted.
EDWARD O. WILSON,
SOCIOBIOLOGY: THE NEW SYNTHESIS

How many people can say that they have spent nine days with a sperm whale? My opportunity came in April 1981 with Physty, a twenty-five-foot, five-year-old male sperm whale nicknamed from an inspired mix of "feisty" and *Physeter macrocephalus.* I was given nine days and nights of the most wondrous, magical, and memorable interaction that I have ever had with any wild dolphin or whale; continuous awe from the moment I saw him, a rare moment when reality was simply overwhelming.

Physty had stranded twice on Long Island, New York, had been towed by a rescue team to a boat basin on Fire Island to recuperate from an unknown illness that had left him floating helplessly, and had endured day after day the feeble but heartfelt efforts of many people who had come to try to make him well. Eleven days after he stranded he swam away on his own, when the time was right. As a volunteer member of a small team of scientists, veterinarians, and an eclectic band of helpers, I was never convinced that physically or medically we did as much for the whale as simple rest accomplished. But we had to try.

Have you ever been to a stranding? Then you know the intense sadness and frustration that people feel when first faced with such a magnificent creature collapsed helpless and vulnerable. Strandings bring out extremes in people, sometimes the worst, but most often the sacrificial, shining best. Every stranding is unique, but there has never been a living sperm whale in a situation like this, before or since.

People helped Physty as they could. For some, that required extraordinary courage. Each person in the water with Physty pushed very real fears aside. No matter what their experience, the teams that came from aquariums and museums were nervous and cautious as they went to work. Our myths about sperm whales include destroying boats and killing people; they do not encourage getting close to one. But the myths came from whale-killing people. Physty's stranding blew the myths away. People relaxed within minutes, as they accepted that the whale was willing. It was obvious that he was aware that we were trying to help. I saw him move or stay still, as if to help, when someone tried something. He gave enormous patience, tolerance, and dignity to our effort.

Few had any clue how to actually help the whale, so filling in around the edges had to do. A wonderful beach-buggy group donated a trailer for meetings and rest. Specialized gear and medicines were volunteered. Coffee and supplies were always provided. Regular visitors became friends. One man exposed his finest available camera and lens to the morning drizzle simply because the experience was worth it, he said. He was thrilled when I helped him get close. One young woman, a professional underwater welder, put her own life on hold for over a week, as did many others. People shared many deep events they had experienced, as if to compare with this moment.

Only two men developed a very personal, hands-on relationship with Physty. Mike, a gentle bear of a man with the touch that quiets horses, was special to Physty the moment they touched. I knew what Mike was when I first saw him put his body between Physty's and a barnacled piling, to smoothly push the whale clear. A professional diver, Mike ran a maritime museum and led the team that stayed for the worst times, and the best.

Jay, the main marine mammal veterinarian, a man with an enormous heart, was hard to miss in his custom-made white wetsuit. He guessed, with much help, what treatment and dosage might make Physty better. All pet vets within miles were startled to have people ask for all the samples of certain antibiotics they had. The plan was to stuff squid with broad-spectrum antibiotics in enormous dosages, to fight immune suppression, pneumonia, and the underlying, unknown illness. Under Jay's direction, an aquarium team tried first to coax Physty to take the medicine-laced squid. Maybe it was the oddness, or touch, or taste, but they failed. Whales and dolphins in captivity must learn to eat dead prey, often only after becoming desperately hungry. And why would Physty bother with such a small sample?

Then Mike was allowed to try. Seemed like only minutes before he was reporting that Physty always sucked the squid in headfirst. Enormous guesswork dosages became possible. One night, because of medical necessity, this most courageous diver attempted to feed Physty squid that had been crammed with antibiotics. With an empathetic affinity for this whale and a red knit cap to help us watch him by the light of car headlights, Mike would take a squid and a breath and drop down to reach Physty's mouth. I was using my Zodiac as the safety boat, staying very, very close as he got one after another squid into the whale. Mike's cap suddenly disappeared out of sight below the dark surface. Just as we were going in to do something desperate, the cap and man popped up. Big smile as he said that Physty had mistakenly grabbed his hand as well as the last squid, rolled a little, and pulled Mike down. Trapped underwater in the dark with his hand clamped in the whale's mouth, Mike did not struggle but rubbed Physty's head with his free hand. Physty immediately opened his mouth, as if embarrassed.

Jay's white wetsuit may have spooked Physty at first. Physty quickly got used to it. On Easter Sunday I made sure all the kids recognized that even the Easter bunny was trying to help Physty. The Jewish veterinarian loved the humor of that image, which I will not let him forget, but he refused to wear a tail and ears for the day.

Say what you want of New Yorkers, but at least 50,000 of them came to witness this one whale, outnumbering all the whalers in history. Enduring rain, wind, traffic, and fumbling explanations, most only saw a still, wrinkled body lying at the surface in a small, semicircular boat basin. People cared a great deal and wanted to know more. Many prayers were said, particularly during the bad periods when the whale's death seemed certain. Physty rested by floating on his side, blowhole exposed, tail curled, looking pathetic and awkward. Even so, Physty obviously recognized certain people. Happily, I was one of those fortunate few. Once, as I walked along the crowd while losing my voice trying to answer their questions, a boy pointed out that Physty had moved along the dock with me. As my clothes changed with the days, I always wondered how he still knew me, but he did. The kids were great, and so were their questions.

Science must be part of this story, but to bring it up is like changing gears without a clutch. From the start I was eager to see what could be learned from Physty. I was concerned that someone would try something invasive, but soon I became astonished that people whose careers were built on this species did not show up, even if just to sit and wonder. Science lost an incredible opportunity with this whale for several reasons. Some experts may have stayed away because they could not deal with an unexpected opportunity that demanded improvisation but offered no controls or privacy. Most expected the whale to die, as all others in similar situations have, and were preparing only for the expected necropsy. Sad to say, some became miffed as their schedules slipped when the whale survived day by day. Plans to move the whale back out toward the sea surfaced too early in the event, almost as if to rid officials of an inconvenience. The schedule was ready, but the whale was not! After our night of frantic calls to attract caring attention, the officials were surprised the next morning to get high-level orders to just help the whale, not sharpen knives. I know there were some who wanted to know how much his brain weighed. Me? Just looking into Physty's eye could lose me. This whale looked back. There was a mind behind that eye. He had a lot of time to wonder what we were, and why we cared. Physty's contemplative stare, with me reflected on his cornea, is a treasured memory of the provocative flood those contemplative interactions brought me.

Most experts who stayed away did so because the working conditions were simply intolerable; people problems kept science from benefiting from this real, live, close-up, and curious whale. Egos, personalities, and autocratic control stifled cooperation and creativity. The conditions were so strained that even a famous *National Geographic* photographer was told to either give up his first two rolls to the person in charge or leave. He left.

Okay, what would I have done, if expert, equipped, and allowed? This was a curious, tolerant, often bored whale. I have no doubt that all sorts of benign, noninvasive experiments could have been started with Physty. I think he would have considered them games; I have seen it happen elsewhere. If you had seen what else he tolerated, you would agree that he would have accepted, perhaps even enjoyed, the fiddling.

For example, sperm whales are masters of sound. This whale made some intriguing noises, often at night, as if in conversation. An array of hydrophones could have been installed around the basin to collect rooms of tapes for later analysis. A smaller array might have been deployed opportunistically around the whale's head from a small boat. Small suction-cup devices might have been attached around his head, if he seemed willing, to record directly from his surface. Physty's acoustical ramblings might have blown away the cloud of ignorance that we still have about the species' abilities with sound, answering questions still debated today. Granted, without some of the more recent computer programs, deciphering the original recordings might have been impossible for a while. Physty's boat basin refuge lacked the ambiance and romance of a sailboat far at sea, but his comments would have saved a lot of today's researchers much guesswork. If the whale had been aware of the recording being made, he might have given us a saga just for the fun of it. You think I'm kidding? Then you haven't been exposed to enough curious wild whales and dolphins, or to the sense of humor I witnessed in this whale. Learning from one of nature's greatest masters of sound could fill books and create careers. Knowing what sounds can be made is only the first step. Knowing how is a giant stride. Knowing why is a leap to understanding. But the ultimate is an interactive exchange between eager and open minds. Physty was ready, but we were not.

Is science ready for the next Physty? People problems aside, there have been all sorts of advances since his visit. Naturally shed skin samples can define a creature's lineage and toxin load. Noninvasive gadgets can scan and probe. Some facts that strandings might provide are very needed, and perhaps not available anywhere else. For example, human noise in the oceans is growing rapidly and rightly assumed to have negative impacts on marine life. But science knows almost nothing about hearing in whales, and, until there is some data, regulations are indefensible guesswork. But how far should we go to get data we need? As one who is always concerned with suffering, I am always judging when the ends justify the means. Judge this one potential experiment for yourself: a Navy-funded team is on standby right now to probe the hearing of any live-stranded whale by measuring neural responses to loud noises and other stimuli, "while there is time." Yes, measuring the hearing of a stressed, dying whale has limited promise for defining the limits of a whole species, but regulators are desperate. The team is eager and well funded. These are the folks who snuck secretly into San Diego's Sea World very early for a few days, before the public arrived, and tried their gadgets with "JJ," the young gray whale later released to the wild. JJ fought the electrodes and wires, and the new gear did not work right. Nothing about hearing in large whales was learned. For a while, Sea World even denied the experiment happened. There will always be a line between what might be learned and the cost in suffering. Only real events and needs will define that line, and it is always changing under pressure. Another event like Physty is certain to include swirling pressures.

Back to the good stuff. I often found us alone together at night. After doing my chores as low profile as possible, I immersed myself in an attempt to understand and interact with this whale, mostly at night when there were no distractions. To my surprise, everyone else went somewhere warm and cozy at night, except some sacrificial students timing Physty's breaths. Art, their enthusiastic and wonderful teacher, helped them endure the wet, cold nights for a deeper learning experience they will never forget.

My car was my bed on wet nights, but whenever possible I was next to the water in a sleeping bag. I was too excited to sleep well, or a close and wet blow often awakened me; a resting whale waited close by, curious or bored. Lying outstretched on the dock, I reached with mind and hand, pondering everything. No, I did not want or need to do anything prohibited by the authorities, threats notwithstanding. In my experience, some individual cetaceans have an interest in figuring us out if given the opportunity. Physty was one, and, even if only in boredom, he interacted with me beyond my perceptual limits. Of course the most vivid moments are the most inexplicable. Isn't that always the way? To someone like me, this was fascinating stuff; I was often immersed in moments that left me happily bewildered. With a reasonably open area of water to move in, Physty instead stayed closer to me than expected, certainly within his body length if the pilings allowed. As the only human there most nights, I seized unique opportunities for unusual data: parking my Honda next to the dock, hatchback open, I played tapes to Physty. Bach brought him straight on and close, the louder the better. He drifted a lot with Vivaldi. So if Physty preferred Bach to Vivaldi, why did he seem so interested in John Denver's singing? Others talked or sang to him too, and I swear that he seemed interested. Somehow my late-night conversations did not seem one-sided, either. Physty often made soft clicks and other sounds when people were close and talking.

Truly enlightening events do not have to flow sequentially. The best moments came or repeated when Physty was feeling well, able to be interested in things around him. A small number of people shared special moments, like having our hands bounced off the small area on front of Physty's head as he clicked. Even at low power, it struck like a blow within a narrow, cone-shaped area in front of his head. Scientists far away were eager to have the exact spot defined for the first time but skeptical of the power we described. Of course no one measured the pulse with some gadget; you had to be there. The rest of his repertoire did not punch like the click did.

How loud could Physty's clicks be? One pre-dawn I jogged around the basin's semicircular dock to warm up. A gunshot echoed in the still, chill air and stopped me cold. With it came a distinct slap against the soles of my sneakers. Physty had stayed where I had left him but had rotated with me as I ran. He was making awesome "chungs," very loud, sharp, echoing, focused, and metallic, never heard before or after. He continued methodically as he searched for the cause of the noise my running made. His pulses traveled over 100 feet in water, carried up the pylons into the dock, and slapped through sneakers to my feet. Incredible power! I continued, he did too, but he stopped when I came close enough to be seen. During the day, usually when there were many people straining to hear our attempts to communicate what we knew about whales, Physty would lift his forehead above the surface and click toward the crowd, but never as loud as that morning. It seemed so obvious that he was trying to understand what was happening around him with controlled sounds.

Chris, a young scientist, even then a renowned expert, did record Physty's sounds for a short period late one night. The big bang never came. Just as well, because the equipment would have been destroyed. As Chris and I sat at dock's edge, Physty moved in, probably curious. Stopped about a meter from the hydrophone, he offered a long sample of grunts, groans, gurgles, squawks, and coughs, but mostly clicks and creaks. We talked and the whale talked and the tape was filled with samples that still confound the scientific literature on sperm whales. Were these social sounds? Was he reacting to any noises we made? We needed help to understand.

We all need help to perceive many things around us that our senses are not equipped to handle. More than sixteen images a second becomes a movie to us. Why not assume that what I heard as a creak was to Physty a specific number of specific sounds? Even a good recording of the clicks Physty made in air might have told us whether he was trying to adapt his clicks to that very different medium.

For one of my more off-the-wall interests, I sorely lacked a gadget to tell me if Physty's ratcheting creak was eight, twelve, or sixteen clicks in very rapid sequence. One night, again close and alone together, I tried the obvious single tap in response to Physty's single click. Then two taps. The whale gave two clicks. Then Physty went to four clicks. After I gave four taps and paused, he made a single rasping sound and waited. Was it eight, sixteen, or . . . what would you have done? Rather than intuitively responding, I was stupidly wondering if Physty wanted a squared number in reply, and even if he used a base often or twelve for numbers, so of course I missed my chance.

Once, later, I tried eight, but it did not get anywhere. Another time, sixteen did not work either. Other people tried the same idea. All of us ran out of ideas, but the whale seemed ready for more. Many of my interactions with wild dolphins or whales over two decades have had a similar result: at some point a really interested dolphin or whale simply leaves me behind. We have started sort of a game, probably with different rules, both interested in the other. The pace or depth picks up. Suddenly I am lost. I call it the threshold, and I have not crossed it yet. However, I do know what a frustrated finback whale looks and sounds like, and the antics of a dolphin that just cannot get the message across to me. I had a few of these moments with Physty as well. It is good to be humbled by one of nature's masterpieces.

Other people may be blessed with some cognitive sense of another's thoughts. I would not label anything I have as beyond intuition. Maybe I am too empirical, or miss subtle cues. But many more moments with Physty gave me much to think about. Once, alone and close together, focused, open, I told him I was "Bill," in a soft voice. I "heard" a sound in reply. I knew it was his name, even if I could not prove it. The unique name that came to me without being uttered, but was at least pronounceable, has served me well ever since.

My nine days with Physty are blurred with emotion, particularly during his ebb times. For a while his breath was fetid. Mucus showed in his blowhole. All of us spent so much time feeling helpless. At low times the pathos was palatable; thousands of people prayed for and willed the whale to get better. Some on the team swam around and even on him with medicine, swabs, and needles, almost in desperation. He endured so many well-meant but painful invasions. For example, because blood samples were demanded, but he had shunted his circulation and no needles could suck any blood out, someone had to cut into his tail to get enough blood flowing for analysis. Always unrestrained, he shivered and flinched but never reacted in any threatening way.

Over the worst hours, we were all sure he was going to be dead any minute. The ebb tide pressed Physty to the bottom on his side. Weak breaths almost never came. A few stood around him in waist-high water; some stroked him, maintaining a vigil that was amplified by the very large, silent crowd. The vigil was a draining honor. I stayed very close but with a strong feeling of peace, some sense of knowing, of friendship. As I became conscious of this, several of us began to talk about what we were feeling, opening up to the moment. I was overwhelmed that others spoke of the same powerful feelings I had.

The entire event magically changed stride with the lifting tide. Physty seemed a different whale, with occasional strong swimming and tail flexing that looked like a yoga routine. The crowd went home with great hopes. That night I tried to watch him closely as he remained active, but the rain and fog finally became too heavy. But he was better!

One vivid memory has never dimmed. Alone and close with the whale on the last night, I simply asked quietly: "How do you feel? What do you want?" He turned, accelerated with three strong pulses of his flukes, and coasted to the net that only symbolically blocked the basin's opening. I never saw him do that any other time. Of course it was coincidental; what do you think I'm suggesting, that this whale understood me and was showing me that he wanted out?

The authorities did want him gone, however. I was joyfully certain that it was the right time for Physty, no matter what the swirl of people planned. The final morning was an officially chaotic operation to herd him back to sea. With several thousand people shouting, "Go, Physty, Go," and in spite of a few things that did not go as planned, Physty was urged out of the boat basin eleven days after he had been towed in. Escorted by a flotilla, he swam deliberately toward open water. He dove deep as soon as the water allowed. No one saw him again. I am as certain that he survived as I am of anything else that happened during my magical time with Physty. When I really let go, I wonder what it would be like to meet him again, now sixty feet of mature grandeur. What stories did he tell of his days with us? I witnessed his sense of humor; I think he laughs about it, however sperm whales laugh.

NAOMI A. ROSE

Sea Change

I grew up in the suburbs of the Midwest, an improbable place to conceive a love for marine mammals. I knew I wanted to be a wildlife biologist by the time I was eleven, but I imagined studying foxes or hawks or some of the other land species I glimpsed rarely but always with a thrill in my wanderings along creek beds and in nearby woods. Yet when I was thirteen, I knew without a doubt that one day I would study cetaceans (whales, dolphins, and porpoises). This conviction arose from a 1975 Jacques Cousteau special. When I watched dolphins playing in the bow wave of *Calypso*, the desire to witness this ocean dance firsthand was overwhelming.

When I was fifteen years old, my family moved to California. The first theme park that I wanted to visit, before Disneyland, before Magic Mountain, was Marineland of the Pacific. This now-defunct marine park had two stars—Orky and Corky, killer whales (now more commonly called orcas) captured years earlier as juveniles from the Pacific Northwest "northern resident" population.[1] I had seen captive bottlenose dolphins and false killer whales, but never live orcas.

I remember noticing at the time that the two whales seemed awfully cozy together in their round tank, the diameter of which was barely twice Orky's length. But I was a teenager—I had not learned to apply critical thinking consistently to the world around me and I did not consider that those cramped quarters were the extent of their universe. I cannot remember if I remarked on Orky's collapsed dorsal fin. I suspect that I thought it was normal for the dorsal fins of male orcas to curl over their backs. While I was thrilled to see these huge creatures up close, breaching and tail slapping, soaking the audience, I remember that ultimately I spent more time watching the dolphins, zooming back and forth in their tanks, leaping high out of the water, reminding me of the wondrous dance I had seen on television at the bow of *Calypso*. Orky and Corky seemed lumbering in comparison, their breaches half-hearted, their movements more labored. I simply accepted that this was how orcas were when compared to their lithe and agile bottlenose dolphin cousins.

I caught glimpses of whales and dolphins off the coast of California over the next few years, but it was not until 1981 that I finally had the opportunity to witness dolphins leaping alongside a ship. It was the summer after my freshman year at Mount Holyoke College, and I spent all three months volunteering aboard the *Atlantis II*, a research vessel owned by the Woods Hole Oceanographic Institution in Cape Cod, Massachusetts. I assisted physical oceanographers with their studies, across the Atlantic Ocean and back. I had not changed my career focus; I just wanted to be at sea. And while I was out there, spending long days cruising across the endless blue of ocean, doing whatever small tasks the oceanographers needed me to do, my dream finally came true. On several occasions common dolphins raced beside the 210-foot *Atlantis II*, twenty to fifty of them at once. Every time I saw the sleek torpedo shapes slicing through the waves, the overwhelming wonder I felt when I first watched the Cousteau special hit me tenfold.

I spent half of my junior year at the University of Hawaii, in a continuing effort to expose myself as much as possible to marine-oriented studies and activities. While there, I volunteered as a student trainer at the Kewalo Basin Marine Laboratory, a research facility that in 1982 had two female bottlenose dolphins, Akeakemai and Phoenix. This long-running project seeks to quantify a dolphin's ability to comprehend and utilize sophisticated language and logic concepts such as syntax and analogy. As at Marineland, I did not pay much attention to the small size of the enclosures in which these dolphins spent their days. I was simply thrilled at being able to touch them. When I returned to Mount Holyoke, I enthusiastically regaled my animal behavior professor with stories of how exciting it had been to observe the speed with which the dolphins learned the artificial languages developed for the project. Although her reception of my tales was polite, one of her remarks struck me forcibly. She said, "Well, that's very interesting, but I would think it would be *more* interesting for us to learn and understand *their* language. Training them to learn a language we made up for them is nice, but it doesn't tell us what *they* think."

While at Kewalo Basin, I was given the opportunity to swim with Akeakemai and Phoenix. The facility had only a very primitive filtration system, allowing algae to grow on the enclosure walls. The volunteers had to scrub them manually, and, to reward us after this arduous task, we were allowed to swim with the dolphins briefly in the newly clean water. Although an experienced dolphin caretaker supervised the interaction, there was a certain element of risk involved, because these two research animals were not trained to be as submissive to humans as marine park dolphins generally are.

As soon as I entered the water, things went wrong—I suspect that the dolphins simply did not like me. They decided to take turns bullying me; one knocked my mask askew with her powerful tail while the other butted me in the ribs with her jaws, leaving me gasping for air. What fascinated me in retrospect (when I calmed down enough to think objectively about it) was that, as soon as I doubled over in distress, both dolphins swam quickly away, appearing subdued and unnerved by my reaction. Had I been a dolphin, I probably would have barely felt the blows—they did not want to injure me but simply to show me who was boss. They had a true social interaction with me, one that they initiated and controlled. It was hardly their fault that I was not dolphin enough to handle it.

Far from diminishing my love for dolphins, this experience gave me a better idea of their complexities. I realized they were not just happy performers or conscientious research subjects, nor were they simply carefree ocean wanderers. They were complex, *profound* animals. They had inner lives. They had likes and dislikes, and moods as well. Above all, they were mysterious; though we could get them to understand us a little, teaching them to recognize certain simple things we wanted of them, we really did not understand them well at all.

Perhaps this mystery is part of the magic that draws people to dolphins. I understand this magic. I can appreciate why so many people want to get close to them in any way possible. I cannot fault this desire, when I feel it so strongly myself. And I have been lucky—I have had the privilege of meeting these creatures in their own world and on their own terms. But along with the privilege, at least for me, came the stripping away of complacency.

My experiences with wild cetaceans forced me to rethink my uncritical and self-centered acceptance of the necessity and value of keeping these animals in captivity.

My epiphany came about over the course of two or three years, starting in 1986, when I first began studying orcas in the wild. When I entered graduate school at the University of California at Santa Cruz, I serendipitously inherited a project studying the same "northern resident" population of orcas from which Orky and Corky had been captured. Every June, on my way up to my field site in Johnstone Strait, British Columbia, my assistants and I would stop at the Vancouver Public Aquarium to confer with colleagues and visit the orcas held there. At that time there were three orcas: one from the Pacific Northwest named Hyak, and two from Iceland named Bjossa and Finna.

Hyak was a huge adult male, and we were usually allowed to observe him and his tankmates more closely than the public could, gaining behind-the-scene privileges through my contacts there. I remember my first impression of his enormous size, which was magnified because his tank was rather small (although not as small as the tiny round tank Orky and Corky had lived in at Marineland). As with Orky and Corky, Hyak and his two friends seemed slow and sedentary, especially compared to the quicksilver white-sided dolphin, White Wings, who shared their enclosure.

After a day or two in Vancouver, I would continue on to Johnstone Strait. My first summer field season introduced me to wild orcas. I witnessed natural behaviors for the first time. I learned about the complex social structure, the tight, lifelong family bonds, and the long-distance travels of these largest of dolphins. I learned that a male orca's dorsal fin normally stands straight and tall, towering five feet or more over his back. The collapsed dorsal fin of captive whales (and of a small percentage of wild whales) is an aberration. I spent long summer days and starry nights in their spectacular coastal environment, with its jagged, rocky shores and wind-whipped waves. I discovered, to my surprise, that orcas are not lumbering and slow, as Orky, Corky, and Hyak had led me to believe. They are swift and powerful and can throw their entire bodies out of the water. They porpoise in the wake of cruise ships and turn on a dime to follow fleeing salmon. They sing eerie, haunting songs, and they can live as long as humans do.

Every time I saw Hyak after that first summer, he seemed bigger and his tank smaller. I started noticing how much time he and his tankmates spent floating motionless at the water's surface, something the Johnstone Strait orcas rarely did. I started wondering how the three Vancouver whales filled their time when the aquarium was closed and all the people had gone home. They did not need to forage and they could not travel anywhere; their tank never presented them with any challenges or changes. I wondered how it must be for Hyak to live with two orcas from an entirely different ocean, and I wondered if all three of them remembered their families. At long last, I was thinking critically.

By the time I finished my doctoral work, I was not so sure about captivity any more. For me, several summers of experiencing the real thing—cetaceans in the wild—stripped captivity down to its basic element: cetaceans trapped in empty concrete boxes. But when I went to work, directly from graduate school, for the Humane Society of the United States as their marine mammal scientist, I learned details about what goes on behind marine park shows that once and forever convinced me that keeping these socially complex, long-lived, intelligent, far-ranging creatures in tanks is wrong.

Orky and Corky were captured in 1968 and 1969, respectively.[2] In my years in Johnstone Strait, I came to know their families. At the Humane Society, I learned that almost an entire generation of juvenile whales had been ruthlessly torn from these families—more than fifty whales were captured through the mid-1970s, out of a population of perhaps 300.[3] A small number had died at the time (from stress or other trauma), and most of the others were dead by the mid-1980s. In contrast, many of the mothers and siblings of the captured animals were still alive in the mid-1980s, and a fair number lived into the 1990s and to this day.[4] Today, only two of the whales captured in the Pacific Northwest survive (Corky, who now lives at Sea World San Diego, and Lolita, who lives at Miami Seaquarium). Some of the Pacific Northwest pods have yet to recover in number, the hole left by the lost generation still unfilled.

No captive orca has ever lived past approximately thirty-five years of age.[5] Both Corky and Lolita are in their mid- or late thirties now; Orky lived to be approximately thirty. All the rest of the captive orcas around the world who have died did so well before their thirtieth birthdays, and many died in their teens. A significant number captured in the 1960s and early 1970s died after only a few months or years in captivity.[6] Yet in the wild, Canadian and American researchers have found that many orcas live into their fourth, fifth, and even sixth decades. Some females are believed to have attained their seventies or eighties.[7] Data from these same researchers demonstrate that the annual mortality rate of wild orcas is about three times *lower* than the rate for captive orcas, a statistic differential that persistently fails to improve with time.[8]

The story for captive bottlenose dolphins is different, possibly because bottlenose dolphins are smaller and more adaptable than orcas and often live in shallow water habitats. As husbandry techniques and facility technology have modernized, the life history statistics for captive bottlenose dolphins have slowly crept up to match those of their wild counterparts.[9] However, this is only true of facilities in the United States and of those that had traded dolphins with these facilities. Facilities in other countries whose husbandry records were likely to be worse were not included in these analyses. Furthermore, it is telling that, in spite of being protected from predators and pollution and being provided with regular veterinary care and reliable meals, captive bottlenose dolphins do not routinely live longer than wild dolphins, whereas these same conditions often allow many captive terrestrial mammals to live longer than they would on average in the wild. In fact, captive bottlenose dolphins rarely approach the maximum potential life span for their species, and the mean life expectancy of a captive dolphin is no better than that of a wild one. At least 50 percent of calves born in captivity die soon after birth. When a captive-born calf dies, a marine park spokesperson will often say that high infant mortality is normal—and so it is, *in the wild*.

But what causes high infant mortality in captivity? What kills captive adult bottlenose dolphins (and other cetacean species) before they reach their maximum potential life span? In a tank, there are no sharks: adults have little to fear from these predators in the wild, but they are a significant source of mortality for calves. There are no food shortages: poor nutrition can lower survival for adults and calves, the latter through affecting the mothers' ability to nurse. There is no pollution: contaminants can weaken an animal's defense against viruses and bacteria. They can also accumulate in a mother's blubber and, through various physiological processes, be eliminated from her tissues through the milk of her first pregnancy, thus harming and even killing her first-born calf.[10] There *is* disease, acquired through tainted fish and possibly through exposure to ill humans,[11] but there is also veterinary care. So, the sources of mortality must be different in captivity, but they are clearly not less severe; if they were, then captive bottlenose dolphins would routinely approach the maximum potential life span of the species.

Proponents of captivity cannot have it both ways: captivity cannot both protect dolphins from the rigors and hazards of the wild and yet be excused from being unable to improve on the species' natural survivorship. Something about captivity clearly harms these ocean creatures. Perhaps constant low-level stress, whether from boredom, restriction of exercise, or lack of choices, weakens captive dolphins' resistance to disease, making them susceptible to infections and parasites that they would normally be able to withstand in the wild. For captive-born calves, I believe that the stress their mothers suffer affects the mothers' physical and even psychological ability to keep their offspring alive.

Little effort, if any, has been made by cetacean capture operators to learn what effects removing juveniles may have on the families and social groups left behind.[12] If an animal is found unsuitable after being restrained and examined and is released, no effort is made to discover his or her fate after swimming off in terror. A study from 1995 of retained dolphins demonstrated that their mortality rate skyrockets sixfold immediately after capture and takes as long as a month to return to normal, strongly indicating that the capture process and subsequent adjustment to captivity is traumatic and harmful.[13] Having watched videotapes of captures, this does not surprise me—the violent process involves nets, ropes, loud noise, extraordinarily rough handling, occasionally the death of companions, and of course an abrupt removal from all that is familiar. Unfortunately, I regularly receive reports indicating that cetacean capture operations are still very much an ongoing phenomenon, not a relic of the past. For example, five orcas were captured in Japan in 1997 for Japanese marine parks—within five months, two were dead.[14] A capture of eight bottlenose dolphins occurred in La Paz, Baja California Sur, in December 2000—within weeks, one was dead.[15] Captive breeding, regardless of species, is still insufficient to replace animals who die or to supply new facilities, so capture operations (in Chinese, Cuban, Japanese, Siberian, and other waters) continue.

People who support captivity because they believe it is educational seem to be aware that something is not right about cetaceans in concrete tanks. Because of subtly expressed discomfort from their otherwise enthusiastic customers (who may be influenced, however subconsciously, by the campaigns of anticaptivity advocates), "state of the art" facilities have built the biggest tanks they can afford, landscaped the above-water surroundings to mimic the animals' natural habitat, and designed the below-water surfaces to look like rock and sand. They continue to work on improving filtration technology so that harmful chemicals are not required to keep the water clean. They no longer house individuals alone, and the shows emphasize the animals' grace, their beauty, and their bond with trainers rather than silly tricks and funny costumes. An ever greater number of small, old-fashioned facilities have gone out of business in the United States and elsewhere. (Unfortunately, the number of facilities is growing in the Carribean and Asia.)

Yet in spite of all these improvements, most captive cetaceans are still held in enclosures that are fundamentally concrete boxes. Their world is conspicuously barren compared to the rich texture of the sea. To me, it appears that the people who support captivity cooperate willingly in their own deception, accepting the architectural illusions as an adequate facsimile of reality. I wonder if their desire to be close to dolphins is so strong that it overwhelms the niggling realization that confining these free spirits is wrong. The very essence that attracts them to these mysterious beings has been confined, modified, controlled, and homogenized in captivity. By packaging the magic and making it readily available, marine parks and aquariums seem to blind visitors (let alone trainers, veterinarians, and curators) to the possibility that the magic has been diluted and even to some extent destroyed.

The most egregious example of packaging the magic is swim-with-the-dolphin (SWTD) programs. I recently participated in such a program as a regular customer, in an effort to improve my understanding of what motivates both the operators of SWTD facilities and their customers. I visited Discovery Cove, a new facility built as an adjunct to Sea World Orlando, just across the street from the larger park. Discovery Cove has between twenty-five and thirty dolphins held in a complex of small holding pens, with at most twelve working in the three larger enclosures during any one twenty-five-minute session. The orientation, a twenty-five-minute talk presented by one of the trainers just before the swim itself, was just what I expected. We had about ten minutes of a simple anatomy lesson, learning about fins and blowholes and skeletal structure. We then had a ten-minute lesson on how to apply basic operant conditioning techniques using positive reinforcement to teach a dolphin how to perform a trick. We heard nothing about dolphin ecology, behavior, social structure, distribution, or conservation status.

The only part of the orientation that surprised me was the last five minutes, when the trainer mentioned that the Marine Mammal Protection Act (the environmental legislation that protects marine mammals under U.S. jurisdiction) prohibits harassment in the wild. He cautioned his listeners not to try to swim with or feed wild dolphins because it might disrupt their normal foraging behavior. There is currently an epidemic of illegal feeding of wild dolphins in certain parts of Florida, a phenomenon some believe is the direct result of people wanting to apply what they witnessed at marine parks to their encounters with wild dolphins.[16] Apparently the government has requested at least some SWTD programs to tag their orientations with the caveat "Don't do this at home, folks." From what I could tell, the impact this admonition had on the audience was minimal. It was a classic case of "Do as I say, not as I do."

The swim experience itself was an eye-opener. It is one thing for young people, untrained in the art of critical thinking, to accept what they see without question. It is another to watch adults, presumably skeptical about any number of things in their lives, allowing themselves to be manipulated because they want so much to believe that their desire to be with a dolphin is reciprocated. When I was volunteering at the Kewalo Basin lab, I recognized that Akeakemai and Phoenix, just as trapped as marine park dolphins but less dominated by their caretakers, saw me as an equal when I was in the water. They made a mental connection with me, in a way not necessarily positive, but nevertheless sincere. At Discovery Cove, Diego, the young male dolphin with whom my group of seven participants interacted, was not connecting with us at all. The difference was very obvious to me, as I spent some time watching where he directed his attention. He was focused entirely on his trainer, watching her carefully no matter what the participants were doing or saying. For her part, the trainer kept up a constant high-energy patter, very similar to a magician who attempts to distract her audience from what her hands are doing so they will not witness the reality behind the trick.

I was most disturbed by the attitude of the trainers toward the animals. Trainers often know little about wild dolphins; they are not necessarily biologists, and some do not even have university degrees. They typically treat the dolphins as they would slightly mischievous but mentally disabled young children, with a bright cheeriness that condescends and has little sensitivity to the potential abilities the dolphins might have. There is no sense that a dolphin in their care is a fully competent, mature being who, in the wild, would be perfectly capable and self-sufficient. Instead there is only a paternal belief that, without the trainer, the dolphin would be helpless. The relationship is one of dominance and submission, however benign. My sense is that the trainer loves her charge but does not respect him.

The trainer controls everything—the food, her demonstrations of affection, when things will start, and when they will end. The dolphin has no choices. Some researchers believe that it is this lack of choice that often leads dolphins (and other captive wildlife) to neuroses, exhibiting the repetitive behaviors referred to as "stereotypy."[17]

If dolphins in SWTD programs really wanted to be with people as much as the people want to be with them, and if the interaction were genuinely spontaneous and mutual, then the risk of a dolphin seriously injuring someone would increase exponentially. (As it is, injuries do occur, albeit infrequently.)[18] This is because dolphins are not gentle, mystical spirits, nor are they simple, benign creatures who are ever willing to do what we want. They are intelligent, emotional, socially sophisticated animals who have a dominance hierarchy and physical ways of maintaining it. As I found with Akeakemai and Phoenix, a human is too puny and weak to withstand a dolphin behaving naturally. No matter what the patter says, there is *nothing* spontaneous about the interactions in a SWTD program. Nevertheless, the people who seek the contact allow themselves to be deceived.

Put plainly, displaying cetaceans in captivity is a business. Very few cetaceans are exhibited sans performance, the way terrestrial species are in zoos. Whether there is a standard show format or a SWTD program, the bottom line is income. (Unlike most zoo animals, cetaceans have considerable economic value; a trained orca can be worth over $1 million.)[19] Even nonprofit facilities need revenue; employees need to be paid and overhead needs to be met. The dolphins are doing a job, one they did not choose. They are paid in fish and the alien affection of humans whom they only understand imperfectly and who do not understand them well at all. They are not tortured, at least not by modern professional trainers, but every aspect of their lives is completely in the control of their caretakers.

The business of cetacean captivity is neither education nor conservation. Claims that it is are illusions and gimmicks. The business is entertainment, as my experience at Discovery Cove made abundantly clear. But the majority of people who now demand that education and conservation be the focus of captive display ignore the fact that the emperor is wearing no clothes. They accept the modern cloak of respectability draping what is still no more than an archaic circus act.

I work every day on marine mammal protection issues, at the legislative, regulatory, and legal level. I see no evidence that the millions of people who visit marine parks every year are any better educated about conservation, or any more aware of environmental issues, than those who do not visit them. I certainly see no evidence that they are *doing* more for conservation. The claim that captive cetaceans are ambassadors for their species simply is not borne out by the evidence. In fact, I believe that marine parks' insistence that the ocean is dangerous for wild dolphins, full of predators, pollution, and parasites, creates a disincentive for people to protect it. If the wild is so inhospitable, surely dolphins are safer and happier in tanks! This is hardly an effective conservation message. The idea that wild cetaceans might actually thrive on the challenge of survival, just as humans do, seems not to have a place in the paternalistic philosophy of captivity.

The societal paradigm I am attempting to shift, with the work I do and through my published writing, is one that is very old and entrenched. It is the paradigm of the menagerie, where the simple desire to see exotic animals up close justifies their domination and confinement. Modern people appear ashamed of this exploitative desire, so they dress it up and cover it with pretensions of education, science, and conservation, but, in the end, it is still exploitation. Lately the exploitation has been of people as well. Rather than teaching people less invasive, more humane, and less dominating ways to channel and redirect the understandable desire for close contact with dolphins, captive display facilities drill home the idea that the *only* way to respond to this desire is through contact with captive animals. This is not education; it is marketing.

I started my career in marine mammal biology not by seeing or touching live animals in a marine park but by watching wild dolphins on television. Therefore I *know* it is not necessary to experience a living animal up close to be inspired to protect it. If the opportunity to experience the magic up close and confined is not available, then people will figure out a way to appreciate the magic from a distance. The paradigm has to be shifted, or individual marine mammals and marine mammal protection will continue to suffer. People leave marine parks with a false sense of security about marine conservation—I know I did when I was young. The glossy façade is so compelling, people become complacent and assume all is well out in the ocean, with the staff at marine parks working hard to protect it. The dolphins are all smiling, after all—and serious and negative news about what is happening in the natural environment does not sell tickets.

I know that if captivity ended, many people would never see a live whale or dolphin. I know that I am suggesting that others must be content not to share in my good fortune. After all, they cannot all go whale- or dolphin-watching. But I am concerned that gratifying our desires, however understandable they may be, has often done tremendous damage to the environment and to wildlife. Perhaps it is time to be selfless, not selfish. Perhaps it is time to try to understand and appreciate the dolphins in their world instead of compelling them to live in ours.

NOTES

1. E. Hoyt, *The Performing Orca: Why the Show Must Stop* (Bath, Engl.: Whale and Dolphin Conservation Society, 1992).

2. Ibid.

3. Ibid.

4. J. K. B. Ford, G. M. Ellis, and K. C. Balcomb, *Killer Whales* (Vancouver: UBC Press, 1994).

5. National Marine Fisheries Service, *Marine Mammal Inventory Report* (Silver Spring, Md.: NMFS, updated periodically).

6. Ibid.

7. P. F. Olesiuk, M. A. Bigg, and G. M. Ellis, "Life History and Population Dynamics of Resident Killer Whales *(Orcinus orca)* in the Coastal Waters of British Columbia and Washington State," *Report of the International Whaling Commission* (special issue) 12 (1990): 209–43.

8. R. J. Small and D. P. DeMaster, "Survival of Five Species of Captive Marine Mammals," *Marine Mammal Science* 11 (1995): 209–26; T. H. Woodley, J. L. Hannah, and D. M. Lavigne, *A Comparison of Survival Rates for Captive and Free-Ranging Bottlenose Dolphins (Tursiops truncatus), Killer Whales (Orcinus orca), and Beluga Whales (Delphinapterus leucas)* (Guelph, Canada: International Marine Mammal Association Technical Report 97–02, 1997).

9. Small and DeMaster, "Survival of Five Species"; Woodley et al., *A Comparison of Survival Rates*.

10. For these last three concepts, see S. Leatherwood and R. R. Reeves, *The Bottlenose Dolphin* (San Diego, Calif.: Academic Press, 1990).

11. M. E. Fowler, *Zoo and Wild Animal Medicine*, 2d ed. (Philadelphia: W. E. Saunders, 1986).

12. Information taken from the conditions established for various capture permits issued by the National Marine Fisheries Service.

13. R. J. Small and D. P. DeMaster, "Acclimation to Captivity: A Quantitative Estimate Based on Survival of Bottlenose Dolphins and California Sea Lions," *Marine Mammal Science* 11 (1995): 510–19.

14. Information taken from action alerts and e-mail correspondence from Japanese animal protection organizations from February to September 2000.

15. Information taken from e-mail correspondence from animal protection organizations worldwide.

16. K. M. Dudzinski, T. G. Frohoff, and T. R. Spradlin, *Wild Dolphin Swim Program Workshop Proceedings*, 13th Biennial Conference on the Biology of Marine Mammals (Maui, Hawaii: Society of Marine Mammalogy, Nov. 28, 1999).

17. H. Markowitz, "Environmental Opportunities and Health Care for Marine Mammals," in *CRC Handbook of Marine Mammal Medicine: Health, Disease, and Rehabilitation*, ed. L. A. Dierauf (Boca Raton, Fla: CRC Press, 1990).

18. N. A. Rose, "To Swim with Dolphins," *HSUS News* (Fall 1994): 11–12.

19. According to memos leaked from Busch Entertainment Corporations in 1993, killer whale purchase prices ranged from $850,000 to $1.2 million.

LINDY WEILGART

Acoustic Smog

Imagine a sperm whale descending into the ocean depths to find food. As she dives from the surface, the sounds of wind and wave recede and soon she is enveloped in an inky, but much quieter, darkness. The water is cold down here, and the sperm whale can feel the enormous pressure of depth squeezing her body and lungs. Still, this inhospitable world is familiar. The great whale is comforted by the echolocation clicks heard around her from family members as they search for squid. They sound like radio static or a herd of horses galloping on cobblestones. She can identify the clicks of an aunt, sister, grandmother, and others who signal success at finding food. Using her own clicks, the sperm whale can pierce the opaqueness of the ocean, illuminating all that is ahead. Squid, her prey, provides very weak echoes because they lack air bladders that effectively reflect sound.

Suddenly, there is a new noise—loud, incomprehensible. It assaults her eardrums, vibrates deep into her floating body like a huge, auditory drill. Disoriented, the sperm whale can no longer hear her family or the faint echoes of her food. Penetrating every fiber of her being, the noise is a terrorizing drone—a predator the sperm whale cannot see or escape. Panic, thrashing, rolling slowly around, unable to navigate. There is just this terrible noise and helplessness. Sound, which used to be the mighty sperm whale's greatest skill and guide, is now her destroyer.

The above is only a guess, though an educated one, of how a marine mammal like the sperm whale may react to man-made undersea noise as the military deploys its new weapon of Low-Frequency Active Sonar (LFAS) in 75 percent of the world's oceans. There have always been natural sources of noise in the ocean, but human-generated noise has increased significantly in the past few decades. Shipping, oil exploration and exploitation (e.g., seismic surveys, drilling), blasting, sonars, and oceanographic and naval projects are all to blame for this rise in noise level. Rather than working toward an abatement of noise pollution, very loud, *new* sources of noise are being added to the din.

Here, I will be focusing on two new sources of noise that scientists and the military have been developing. The Acoustic Thermometry of Ocean Climate (ATOC) project was first designed to measure ocean temperature by sending very loud sounds across whole oceans. ATOC has since been renamed North Pacific Acoustic Laboratory (NPAL) and plans to operate off Hawaii for periods of years. The U.S. Navy's LFAS program uses extremely loud sonar intended to detect newer, quieter enemy submarines. The ATOC/ NPAL and LFAS projects present a long-term, global threat involving large parts of the oceans of the world, and their necessity to the world has not been convincingly established.

Over the past twenty years, my husband, my three young children, and I have lived among sperm whales for months at a time, studying whale sounds and communication. At times, we have shared their offshore environment more than that of our own species. We have witnessed the birth of one of their young, and I have nursed my babies on deck while they have nursed theirs. I have repeatedly witnessed the extreme sensitivity sperm whales show toward noise. Once, while I was observing sperm whales from the mast of our sailboat, one of our crew entered the water. I saw the swimmer, tiny against the massive presence of the surrounding whales, make one careless splash with his legs, and suddenly, almost laughably, all the whales panicked and disappeared in haste. It was obvious to me, watching from above, that the whales were aware of the swimmer's presence and only reacted to the sound. However, I have managed to follow closely groups of adults with calves by swimming very quietly without flippers, using a splash-free breast-stroke. Some of the most memorable moments of my life have occurred while swimming with sperm whales, immersed in the patterns of clicks, or "codas," they make to communicate with each other while at the surface. At close distances I could sense, rather than just hear, these clicks drumming and pulsating through my body.

It is difficult for me to imagine how the long-term use of LFAS or ATOC/NPAL could fail to change the lives of these whales who are so sensitive to underwater sound. Past scientific studies on the effects of noise on whales have indicated, with remarkable agreement, that at least two species of whale consistently avoided underwater noise at an intensity level (at their ears) of around 120 decibels (dB; all levels are given as water standard, because decibel levels underwater are not comparable with those in air). ATOC/NPAL sounds begin at the source at 195 dB and drop to 120 dB about thirty miles away from the sound source.[1] LFAS, with a source level of 230–50 dB, drops to 120 dB over 300 miles away, so that any whales or dolphins swimming within this area—an area greater than the size of Texas—could be affected by noise louder than the 120 dB known to disturb them.[2] I am concerned that we are creating an "acoustic smog" that will interfere with whales' ability to find mates over long distances (which would affect reproduction), listen to the quiet sounds emitted by predators and prey, detect beaches and avoid stranding themselves or getting tangled in fishing gear, stay with members of the group, and stay in contact with their calves.

Under the roughly $43 million ATOC project, scientists at Scripps Institute of Oceanography first proposed to place a total of ten to fifteen underwater speakers in all oceans and transmit very loud sounds around the world for a decade or more. The supposed purpose was to use sound to gain information on global warming, because sound travels faster in warmer water. The first two speakers were to be located off Kauai, Hawaii, and in the Monterey Bay National Marine Sanctuary off California, both biologically important and rich areas. Public outcry in 1994 forced Scripps to curtail its plans and move its California speaker out of, but still near, the sanctuary. ATOC was allowed to proceed for about three years of broadcasts. Very loud sounds broadcast off California were received off New Zealand, over 6,500 miles away. NPAL still uses the underwater speaker off Kauai.

Marine mammal scientists were funded by ATOC to carry out studies in an attempt to determine whether ATOC was safe for marine life. One study showed that sperm and humpback whales clearly avoided the speaker while it was transmitting (compared with when it was off). However, these results were deemed "biologically insignificant" by the ATOC marine mammal scientists and were brushed aside. NPAL received authorization to proceed on January 15, 2002, from the National Marine Fisheries Service, the governmental regulating agency. NPAL is authorized to operate for five years but is reviewed every year. NPAL has receded from the spotlight somewhat as public attention has shifted to the LFAS project.

As with ATOC, in 1997 and 1998 studies were designed and carried out in conjunction with the $350 million LFAS project to determine how four species of whales would respond to the sound. During test broadcasts of LFAS off California, scientists observed a decrease in the number of calling blue and finback whales and a dramatically clear avoidance of LFAS for inshore migrating grey whales. On the breeding grounds in Hawaii, half of the studied humpbacks stopped singing, even at great distance from the sound source. Humpback songs also became significantly longer, perhaps to compensate for the noisy interference. Such reactions were again judged "biologically insignificant" even though trials used a much reduced (quieter) power level for the sonar than would be the case during actual deployment. Marine mammalogist Dr. Marsha Green of Albright College and the Ocean Mammal Institute wrote that LFAS was tested at low levels on only four species of whales for about one month each. She concluded that we know virtually nothing about what impact the higher, deployment-level sonar will have on marine life and humans over the long term.[3] LFAS was granted authorization to proceed on August 16, 2002, for a period of five years, reviewed every year. However, the Natural Resources Defense Council, along with other environmental and animal protection groups, sued the National Marine Fisheries Service for granting LFAS authorization. On October 31, 2002, a federal judge concluded that LFAS violated several federal environmental laws and ordered the U.S. Navy to work with environmental groups to find a compromise. Post–September 11, this decision against the military was especially noteworthy. The judge was apparently unconvinced that the whale studies carried out in conjunction with LFAS demonstrated the project's lack of danger to marine life.

The marine environment is difficult to study because we are terrestrial beings and most of the ocean is opaque to us. Oceans are variable and change from season to season and year to year for unknown reasons. Our inability to manage fisheries stems from these problems (in addition to our greed). Whales and dolphins are hard to study because we see so little of their lives at the surface, let alone underwater. They are also long-lived and slow reproducers, which means that they are more threatened by environmental degradation. We do not have even remotely accurate counts for most whale and dolphin populations; various estimates for the same population can be off by a factor of ten or more. How can we hope to assess the impact of noise on a given whale population if we do not even know how many animals there are to start with, not to mention what their birth rates, death rates, and growth rates are?

Yet these statistics are required to assess whether noise is threatening a population's survival. Marine mammal studies such as those done as part of the ATOC and LFAS projects are interesting, but they do not come close to determining whether a particular noise is safe or not. We cannot conclude that an affected whale can still find as many mates (some whale species may communicate by sound over thousands of miles to contact mates), as much food (whales may need to listen to the faint sounds their prey makes to find concentrations of food), avoid predators, avoid hazards like beaches and fishing gear, navigate, and stay in contact with their group and their young, all of which is done using sound. Interference in any of these activities of daily life could have serious consequences (eventually even extinction) for the population.

And how can we assess hearing damage in the large whales? If whales show less and less reaction to noise over time, does this mean they are getting used to it or going deaf? And just because some dolphins may subject themselves to high-noise conditions, can we assume they are not being harmed? Maybe they have to put up with a noisy environment to get access to food. We know sound is damaging well before it becomes annoying to us. Even in humans, the assessment of damage from noise is controversial. It is difficult to define objectively those characteristics of noise that make it irritating to us. There is also great variability between human individuals in their susceptibility and sensitivity to noise. We can only imagine the variability in sensitivity that exists between different whale and dolphin species, age classes, individuals, and sexes, as well as between animals engaged in different activities or in different areas.

To determine the health of a population and its survival, we need detailed, accurate population size figures, birth rates, death rates, and growth rates. We have none of these statistics for almost any whale or dolphin population likely to be affected by LFAS or ATOC/NPAL. We do not even really know which sounds the great whales can hear. In addition, evidence suggests that exposure to loud sound not only disturbs whales and dolphins but can also be deadly. On May 12–13, 1996, twelve Cuvier's beaked whales stranded and later died along the Kyparissiakos Gulf coast of Greece. These deep-diving, small, toothed whales very rarely mass strand. Moreover, this stranding was atypical in that the stranded animals were spread out over a large, thirty-eight-kilometer coastal area. Such stranding characteristics would fit with a cause that had a sudden onset yet affected a large area of ocean simultaneously, such as a loud, underwater sound. NATO's LFAS was tested in the area beginning on May 11, one day before the whales stranded. The timing and locations of the strandings and LFAS broadcasts were almost perfectly coincident with one another. Alexandros Frantzis, a Greek scientist who reported the findings in *Nature*, a British scientific journal, concluded in his article, "Although pure coincidence cannot be excluded, it seems improbable that the two events were independent."[4]

Some scientists think the stranding was a result of a panic reaction by the whales, but this is only an after-the-fact theory. No scientist had predicted this exact reaction; they could not, given our profound ignorance of the effects of sound on marine mammals. However, many of us had felt that deep-diving, toothed whales were most at risk for being harmed by acoustic pollution because they are adapted to much quieter conditions.

Since the 1996 Greek stranding, even more evidence of the deadly effect of sound has come to light. On March 15, 2000, another atypical stranding occurred, this time in the Bahamas. Seventeen cetaceans, including little-known species of beaked whales, minke whales, and an Atlantic spotted dolphin, beached in the Bahamas at the same time as a fleet of U.S. naval warships using mid-range tactical sonar traversed the area. An interim report released by the U.S. Navy and the National Oceanic and Atmospheric Administration (NOAA) on December 20, 2001, stated that seven whales died. Moreover, it noted that samples of the dead whales collected by Ken Balcomb of the Bahamas Marine Mammal Survey showed hemorrhaging in the brain, in and around the inner ears, and in the jaws (an important structure for hearing in whales). The report concluded that the most plausible cause for the stranding was exposure to intense sounds, specifically the Navy's active sonar.

Prior to the naval activities, Balcomb had been studying a group of beaked whales in the area over a period of nine years. Since the stranding, only one of the thirty-five whales in his study has been seen again. He believes that most, if not all, of the local population of this species was killed by the naval activities in "an acoustic holocaust that can be likened to fishing with dynamite."[5] At the very least, he concludes, there was very serious displacement of these whales.

We have no idea how many such strandings go unwitnessed or how many whales and dolphins die at sea as a result of being assaulted by noise. Panic reactions could conceivably cause whales or dolphins to surface too quickly from depth, for instance, with grave consequences. It is important to bear in mind that the two cases noted above were only discovered because a biologist was present and because they did not occur in remote locations. Therefore, these strandings in all likelihood represent only a tiny fraction of the total strandings or deaths at sea that may be occurring as a result of noise pollution.

What can we learn from these strandings so that these whales and dolphins will not have died in vain? I would argue that the strandings should serve as a strong warning against our invasions into nature when we have no inkling of the consequences of such actions. First and foremost, we must humbly admit that we know almost nothing about how whales and dolphins perceive, react to, and are affected by noise. Moreover (and this is a tough one for many to accept), we are, despite our best attempts, unlikely ever to have a thorough and complete understanding. The question "What is the effect of undersea noise on whales and dolphins?" will remain, for all intents and purposes, unanswerable. Should we then abandon all scientific inquiry into the subject? No, we should continue with studies into noise pollution, provided that these studies (1) are not used as a delay tactic in the regulation of undersea noise, (2) do not claim to answer the above unanswerable question, (3) are not open to misuse so that the inability to find a detectable, harmful effect of noise is not interpreted as proof that the noise is safe, (4) do not themselves contribute to the addition of further noise into the marine environment, and (5) are not *directly* funded by noise polluters, resulting in a conflict-of-interest situation.

Although the preceding question is largely unanswerable, another question—"How can we make undersea noise less harmful?"—*is* answerable. This falls under the category of risk reduction in the face of uncertainty. We can look at the range of possible outcomes of our actions and make rational decisions on how to reduce bad outcomes. This may mean making heavy use of the Precautionary Principle at times, where measures are taken to prevent harm even in the absence of conclusive scientific evidence, or erring on the side of caution.

Simply eliminating or reducing "acoustic smog" is probably the safest course to pursue, though it may be the most unpopular with polluters. Nevertheless, it is achievable and does not need to be as painful as it sounds. To those who say I am being impractical, I would argue that bringing about the use of double-hulled tankers sounded impractical at one time, too. Although retrofitting ship engines to make them quieter is expensive, building quieter engines in new ships is not. (The U.S. Navy has the technology and is well versed in making quiet engines.) Simply maintaining ship engines reduces noise substantially as well as making them more efficient. Engines are also much quieter at slower speeds. Shipboard workers would share in the benefit of quieter engines by enjoying quieter living conditions onboard. Results from seismic surveys could be shared among oil companies so that the same section of ocean would not need to be repeatedly surveyed by each company. Finally, oceanographic or naval projects that operate over vast scales of space and time should be strongly curtailed or, preferably, eliminated. When elimination is not possible, altering the locations and timing of noise sources to avoid areas and times of year in which marine life is abundant greatly reduces the risk of impact. Critical habitats or "hot spots" are easily identified, and this approach alone would go far in protecting whales and dolphins. It is particularly unfortunate that a stationary source like NPAL is located in a biologically sensitive area close to a humpback whale sanctuary and an area used by endangered breeding monk seals.

Many concerned scientists internationally believe the LFAS and NPAL programs should be halted. There are far too many critical issues to address before the Navy or NPAL goes forward with use of their technologies. No one has even studied the ecological impact of using LFAS or NPAL. So far, the Navy, though finally admitting its probable responsibility for the Bahamas stranding, has failed abysmally to provide any assurance that this tragedy will not be repeated—nor can it, given our meager knowledge. And sure enough, on September 24, 2002, sixteen beaked whales of three different species stranded on the Canary Islands. As of this writing, nine whales had died, and severe hemorrhaging in the brains of the otherwise healthy animals had been discovered. NATO maneuvers using underwater sounds were conducted at the time and in the area. At least fifty-eight ships, six submarines, and thirty airplanes participated in the exercises. How many more whales need to die a senseless death before we take action?

As the Navy rolls out its newest military weapon, our oceans will be dying the slow "death of a thousand cuts"—a phrase Dr. Sylvia Earle, former chief scientist at NOAA, used to describe undersea noise pollution.[6] This underwater assault will go unheard by humans, but that does not mean it is not happening. The oceans give us our air, water, and food, and they regulate our fragile climate. The oceans are literally our life source.

As a scientist—and as a mother and fellow inhabitant of this fragile planet—I am alarmed at these new threats to our oceans. I cannot imagine why we would subject our seas and marine animals—many of whose lives depend on sound—to yet another source of acoustic pollution. As we are learning with nuclear weapons, there are some technologies that simply should never be used.

NOTES

1. Office of Naval Research, Final Environmental Impact Statement for the North Pacific Acoustic Laboratory, vol. 1 (May 2001), 2–13.

2. J. Potter, pers. comm.

3. Marsha Green, testimony at National Marine Fisheries Service public hearing, Silver Spring, Md., May 3, 2001.

4. Alexandros Frantzis, "Does Acoustic Testing Strand Whales?" *Nature* 392 (1998): 29.

5. Ken Balcomb, letter to Joe Johnson, U.S. Navy, Feb. 23, 2001.

6. Sylvia Earle, "Defining the Problem," in Michael Jasny, *Sounding the Depths* (New York: Natural Resources Defense Council, 1999), vii.

The Discovery
and Development
of Dolphin-assisted Therapy

M any important scientific discoveries are made not by deliberate design but as by-products of some individual scientist's curiosity, and I admit my discovery of the relationship between the handicapped and dolphins was such serendipity.

In 1971 I found myself teaching in Miami, where I read a brief newspaper account of a research center called the Dolphin Project. I immediately decided this was an opportunity to study human-dolphin communication. It was through this contact that I discovered the passionate interest to which I would dedicate myself.

In the winter of 1972, my brother came to visit. David was four years younger than me, and a serious childhood illness had left him neurologically impaired, classified as mentally disabled. I was delighted to take him to see the project's two dolphins and, while at the site, decided to let him go into the water with Florida and Liberty.

Both dolphins, one male and one female, were then adolescents, quite assertive and spoiled by the many people who catered to their slightest demands. I did not expect David to get very close to them, but he immediately walked into the water up to his waist. That was when an extraordinary event occurred. Liberty, the rambunctious male, came toward David at full speed but stopped quickly when David spoke to him in a quiet tone. Liberty became gentle and attentive and remained still as David stroked him and slowly cupped water over his body. Florida joined them, and David stroked both of them while they swam patiently around him.

I was stunned! These dolphins had never stayed still for a second! I called

David to come out of the water and then had him return. The same thing happened with the same results. Then I removed David from the water and went in myself. Immediately Liberty rushed and pushed me into the deeper water, and Florida joined the game by trying to force me under the water to play.

Had they reacted so differently with my brother because he was neurologically impaired? I had a specific question for my interaction research.

It was not until 1978 that I again turned my attention to dolphins, with the establishment of project INREACH at the Miami Seaquarium. The project had two primary goals: to see if an affinity exists between the children and the dolphins; and to establish if interaction between them would cause any behavioral or verbalization changes. Dr. Henry Truby, Nancy Phillips, and I established simple criteria for program participation: the parents must understand the project purposes and agree that they and their children would commit to these goals. The children were to be old enough to have "language" but not demonstrate "speech" in their expressive vocabulary. They should also not be frightened by the water or animals. Eight children between the ages of ten and seventeen were selected to participate. The majority of the children were classically autistic, and the rest suffered neurological damage from birth trauma or unspecified diagnoses.

Six encounter sessions, each from four to six hours, were held between December 1978 and August 1979. An important development during project sessions was the consistent increase of sustained attention span during dolphin-child encounters. Children who before had reported attention spans of five to ten minutes now reported sustained attention spans of up to one hour.

Michael Williams, a fifteen-year-old diagnosed at age six as "nonverbal autistic," exhibited the most dramatic observable behavioral differences, including an interaction where he and another autistic child each held a side of a bucket, lifted it together, and, in a simultaneous movement, poured the water over a dolphin. Neither adolescent had ever demonstrated interactive, mutually cooperative play previously.

I knew I had again discovered intriguing and unanswered questions. Why do dolphins respond so intensely to handicapped people? What triggered the cooperative bucket activity? Once the children are placed in the water with the dolphins, will tactile-aural situations create major avenues for further communicatory attempts by the dolphins or the children?

This last question initiated a mutually agreed upon separation between Dr. Truby and me. Although it was clear to both of us that the children needed to be placed in the water with the dolphins, I did not believe that chlorinated cement tanks were an appropriate milieu. After an amicable parting, Dr. Truby decided to discontinue his research in this area.

In 1981, two years later, I read an article in a local newspaper about a family in Key Largo that had captured six Atlantic bottlenose dolphins to establish a captive-display business called Dolphins Plus. The six untrained dolphins living in the Atlantic Ocean side of Key Largo appeared optimum for my research. The Borguss family accepted my proposal to conduct research, and I contacted the Williams family. They had been so favorably impressed with their son Michael's progress during the earlier project IN-REACH that they readily volunteered to be the first case study.

Michael was now seventeen years old, almost six feet tall, and very huskily built. He was completely nonverbal, with self-abusive behaviors occasionally turning on other people. He spent hours self-stimulating with excessive arm and hand flapping as well as occasional circle whirling. He had an inordinate fear of motion and would not use escalators or stand on any elevated or extended platforms or walkways. He did not like new situations and rejected changes in his routine. He also had a severe sleep disorder that made him unable to sleep the night through. He showed little emotion except to display seemingly appropriate responses to cues taught by behavioral training techniques.

Progress was made over the years as Michael and I worked with the dolphins. He no longer displayed abusive behavior. His attention span increased to the point where he could concentrate on daily living skills for well over twenty minutes at a time. He acknowledged people and things in his environment, and his parents were able to take him along shopping and visiting. He lost his fear of open spaces and did not cling to the walls of buildings. His terror of escalators was eliminated, and he moved freely in the world around him.

But everything was not idyllic during these years; Dolphins Plus's financial situation remained extremely precarious. Without the dolphins trained to perform in "shows," income was inadequate, and the Borguss family had to maintain outside employment to meet their expenses. One fateful day, Rick Borguss informed me that he had arranged to sell two of the dolphins, George and Angel, to a European dolphinarium. I was stunned by this news and brought face to face with the sobering market realities of the business world. George had patiently taught me how to behave around dolphins, and Angel had led the therapy sessions, but now, after three years and a bond with these dolphins impossible to put into words, I faced their loss. The thought of their spending the rest of their lives as performing clowns in chlorinated tanks caused great pain. I faced for the first time the full consequential reality of removing dolphins from the wild for profit. In August 1984, the two dolphins were transported to Europe. (Angel is still performing; George died in 1989.)

In 1984 Dolphins Plus opened the first "swim-with-the-dolphins" program in the world. Rick concluded that, if the dolphins treated the disabled so carefully, then certain aquatically oriented people could also swim with them.

As more people came to the site, I established observation studies with other disabled people and examined how they interacted with the dolphins. Adults and children with various physical and mental disorders participated, and this permitted me to study dolphin behavior and perhaps discover the limits of dolphin influence. Yet I never found that limit, being constantly amazed at the dolphins' focus and adaptability to people's needs. Even when it was questionable if the experimental goals had been met, something unplanned would occur with positive effects for those involved.

It was now clear that a controlled experimental situation was indicated. Setting up controlled experiments with autistic children and dolphins is not simple. A semblance of control demanded a twenty-four-hour live-in situation for both children and support staff. While half of the children were working with the dolphins, the other half absolutely needed a complementary schedule of water activities.

I discussed my preliminary experimental parameters with Rick, who agreed Dolphins Plus would close for the project duration and suffer the financial loss if I put the remaining pieces together. Our research protocol was set, and the goals of the "Dolphins Plus Autism" project came into focus.

Nonverbal adolescent boys considered seriously socially and behaviorally impaired were selected, and target outcomes would be an improvement of the dolphin group's appropriate interaction skills. The long-range goal of these target outcomes would be sustained changes over time in nondolphin contexts.

In the Florida Keys, July is hot and humid with daily temperatures in the high 90s F. Engaging in strenuous physical activity over a prolonged period took its toll on us in both the experimental and the control groups. During this time of year, heat also affects dolphins, who tend to slow down and interact more sporadically and leisurely. During the entire project time, however, the dolphins were the most responsive and attentive I had ever seen. After spending hours at a time in the water, the humans sometimes seemed to wilt, but the dolphins appeared progressively energized by these contacts.

We had two work platforms. One of them, used by Julie Baxter, an occupational therapist, was for the "orientation" situation, where Julie introduced the children to the dolphins and to each other through play instruments (rings, balls, etc.). After achieving a degree of comfort, each child was transferred to the second platform where Rick and I pursued in-water interaction with the dolphins.

We used fish at first as a reward in each session, but I noticed that the dolphins increasingly interacted before the fish bucket arrived; some dolphins even preferred the children on the platform where no fish were available.

We stopped using a food reward for in-water contacts, and, to our surprise, it made no difference in the dolphins' eagerness to play and tow. The orientation platform cut the fish allocation in half, without affecting platform play behavior.

I then made a fundamental observation: the dolphins were not attending to the fish or our directions; they were instead following individual children. An attachment had been formed to a specific child, and no matter what platform that child was on, the dolphin was waiting there.

I also discovered that we were not alone in directing activities; Dingy, the alpha female, would give commands to the dolphins at critical times when certain actions were demanded. I noticed she helped another dolphin when we were physically prodding a child, and she would round up two or three of the dolphins for a conference and then disperse them to specific platforms. Sometimes L.B., the alpha male, would get overly excited while playing, would move too quickly past a child or throw a ball too hard to catch, or just race back and forth. Dingy would come alongside, synchronize her swimming, and slowly reduce their speed until he became calm.

Two things were now clear: never again would a fish reward be used, and the dolphins could choose with whom they bonded during working sessions. The dolphins were not looking to perform a task and receive a reward; they were actually involved in the process, working not *for* us but *with* us.

Data analysis from the controlled setting showed no significant statistical differences (pre/post) for our small sample size (seven subjects). The only clear trend was, given the choice of structured activity (as opposed to self-stimulation) or nothing, the dolphin group always chose activity.

At our six-month post meeting, the questionnaires returned by the parents showed some interesting responses. All stated that the children were calmer and more self-sufficient. The experience of being away from home, having to adjust to daily situations, and the confidence engendered was carried through in home behavior. Children in the dolphin group with sleep disorders showed marked improvement in sleep patterns that were maintained over the six-month period.

Statistical evidence about dolphin-assisted therapy was not overwhelming, yet information and knowledge gained led to dolphin-assisted therapy restructuring and represented a quantum leap in my understanding of the dolphin's role in the process. For the next five years, I continued to conduct exploratory studies with the blind and deaf and others with physical disabilities, as well as those with mental and emotional disorders. Even my athletic, seemingly indestructible husband became a research subject after he suffered two cerebral hemorrhages that required five neurosurgeries and six months in the hospital. The five-year odyssey of rehabilitation took dolphin work from the academic world and placed it squarely in my personal life.

From the beginning of my research, my husband had a wonderful relationship with the dolphins, swimming for hours up and down the canal that opened to the ocean. After release from the hospital, we brought him to the dolphins, still in a wheelchair, which required him to be lifted into the water. Dingy and L.B. were at the time eagerly swimming close to the platform, and when Arthur and I entered the water, the dolphins immediately realized something about him was different.

They would crowd him before, bumping, pushing, and eager to dive and sprint through the water. They now came to his disabled right side and "sounded" the length of his body with their echolocation as they swam very slowly alongside. Dingy gently rubbed along his right side and lifted him slightly as we swam. Even my optimistic and cheerful husband had some dark moments in his long and arduous rehabilitation. A trip to the dolphins always changed a negative mood and restored his positive spirit of recovery.

I expanded my contact with wild dolphins during this period by swimming with pods in Hawaii, the Bahamas, and Japan. I swam with boisterous spinner dolphins who filled the water with staccato chatter, and I peacefully floated with a serene spotted dolphin who brought me her baby to admire. In Japan, small graceful bottlenose dolphins, familiar with only a few island people, demonstrated an unguarded openness that captive dolphins, anticipating the next human demand, never know. I eventually stopped all work with captive dolphins.

My reasons for no longer working with captive dolphins are multiple. First, upon return from wild dolphin swimming, it became difficult to see Dingy and L.B. confined, and I found myself thinking often of George, who died prematurely in a cement tank, and Angel, condemned to captivity until she dies. I had learned much from them and could do nothing to reciprocate.

Second, dolphinariums and swim programs are for-profit businesses, often charging up to $100 to swim with the dolphins. I questioned removing sentient beings from their natural habitat for the sole purpose of financial gain. Dolphin businesses will often justify their exploitation under the therapy pretext. Every dolphinarium and swim program extolled the "therapeutic value" of dolphin contact, even though a child having fun is not equivalent to therapy. It is a rather cynical and deceptive practice by dolphinarium and swim-program owners. Some certified therapists with no dolphin knowledge will charge exorbitant fees for treatment that can be done without dolphins. Often so-called "self-help" people put together package programs using the dolphin swim experience as an excuse to double or triple their fees. At the heart of all these therapy programs is the exploitation of vulnerable people and vulnerable dolphins. My name or references to my work were often associated with the public-relations come-on appeal extolling the virtues of these programs. Therefore, I had to stop all association with captive dolphins.

During all the years my colleagues and I developed dolphin-assisted therapy, we never charged anyone and asked people only to give us their trust and to open their lives to us. In turn, we offered not just our professional skills but also our minds and hearts. Every therapist with me was licensed and had exceptional professional skills. This special situation created by the dolphins demanded they open their lives to the totality of the experience. It never occurred to anyone that a fee for research activities would be charged.

What about the dolphins? Those I worked with were classified for display and controlled by others, and I will always feel some responsibility for what happened to them. I believed that turning people's attention to wild free dolphins could present a possible alternative. I hoped the interaction with open ocean dolphins would expand the parameters of dolphin-assisted therapy, and I conducted a case study as part of a teaching program. The therapists and I explored the various venues available for projects with wild dolphins, but we rejected site after site for ongoing research because of the dangers presented by people.

In the rush for personal pleasure, people disregarded the damage that could be done to the other species, and swimming with wild dolphins has now exploded into a worldwide business. Those studying wild dolphins report humans chasing them with boats and polluting their water with noise and garbage, creating very stressful situations. Therapeutic purposes are often the justification given for this rude invasion.

People profiting from dolphin-assisted therapy do not wish to assemble or sponsor double-blind studies needed to test my exploratory research, and no scientific evidence exists that dolphin-assisted therapy is more effective than traditional or other adjunct therapies. Still, I have no doubt that those individuals we worked with over a ten-year period where greatly helped by the dolphins' assistance. It was the integration of talented therapists, involved families, and my personal commitment to giving each individual dolphin respect that created our many small miracles.

However, the question that became clear was whether what we did justified imprisoning dolphins. I had to consider ending my research after assuring myself that there were many other fine therapies, including work with domesticated animals, providing the same success stories. After a bout of deep personal angst, I decided to stop all dolphin-assisted therapy research in 1992. Perhaps it is time for us to leave the dolphins alone.

LEIGH CALVEZ

Dolphin Lessons

T he climb was treacherous at best, but this was the perfect spot to observe wild, Hawaiian spinner dolphins. If I carefully placed each foot in the loose lava rock and found secure handholds among razor grass and dead trees, I could make it to the site. I heard the scuffing and then sliding of miniature avalanches created by the members of the research team climbing ahead of me. As heat radiated up from the rocks and the sun beat down on my back, I made my way slowly up the hill for the last time. I remembered it was much cooler here in the summer—the rainy season—when a layer of clouds formed each afternoon, blocking the intensity of the sun. At last, I reached my destination, a precarious precipice, shared with three or four other researchers, 150 feet up overlooking the bay. I looked out over the clear water—shades blending from sandy brown to aquamarine to lapis—searching.

It was February 1999, and I was leading a research project for the Ocean Mammal Institute to study the effects of swimmers, kayaks, and motorboats on pods of spinner dolphins on the Big Island of Hawaii. We studied dolphins and humans from our cliff site using binoculars and a theodolite (a surveyor's tool used to mark location) to follow movements and behaviors. Each day we ascended the nearly vertical cliff, balanced ourselves and our equipment on the tiny ledge, and watched the exchanges between people and dolphins in the bay below. We worked in three-hour shifts between 7:00 A.M. and 4:00 P.M. With the perspective we gained from our elevated station and the aid of binoculars, the dolphins were clearly visible when a group of tiny, gray dorsal fins broke the surface.

On this last day of research, we climbed up the cliff for a midday shift. I perched on a raggedy chair in the tiny patch of shade provided by the tarp and pulled a pair of binoculars out of my backpack. I scanned the bay looking for dolphins. The view of the bay from the cliff reminded me of watching an event from the top row of bleachers in a stadium, although not as comfortable. I spotted the dolphins amid a circle of kayaks. This bay is very popular both with dolphins and with people who hope to swim with them. The inlet is sheltered, the water usually free from threatening currents, and it is one of the most reliable places to find spinner dolphins. Over two months of observation, I had learned to tell the time of day by watching the interspecies interactions.

As we prepared ourselves for our day of research by arranging the equipment on our tiny ledge and applying oily sunscreen, I tried to imagine the pod's long night out at sea searching for enough food to feed each member. I imagined the dolphins surrounded by the cold blackness of the ocean, using their sixth sense of echolocation, a natural type of sonar, to navigate and find food. Click . . . Click . . . ClickClickClick. Pow! The small squid floated, stunned by the sound, as a hungry dolphin quickly sucked up this bite of her evening meal. The pod continued dining at depth on the school of squid. It was a large amount of food, and they were lucky to find it. They had not been so fortunate the night before and went hungry. The dolphins ate their fill and more, not knowing when they would eat again.

As the pod finished its meal, the dolphins made the decision to travel back to shore to spend the day resting in one of their favorite bays—a shallow area with sand covering most of the bottom, which meant easy detection of predators like the tiger shark. The pod had moved far offshore and would have to travel about eight miles southwest to reach its preferred destination. As the spinners neared the area at dawn, they heard sounds in the water coming from the bay. They used their echolocation, sending out a sound pulse to determine the source of the noise. The echoes they received were familiar—humans. I tried to imagine what the dolphins would think of this discovery. *Oh humans!* a dolphin might note excitedly as she dashed off to greet them. *Not humans again!* another dolphin might think, and move away from any interaction.

I shifted in my chair on the slender ledge high above the bay and adjusted my binoculars as I thought about what the dolphins faced each day when they came inshore to rest. Early in the morning, from about 7:00 to 9:00, there are few people swimming. The locals who swim with dolphins and those with the most experience usually swim at this time. Normally the dolphins have just come into the bay for a day's or night's rest, depending on one's perspective. During these hours, when the pod is playful, the dolphins leap and spin, fill the water with chatter, and seemingly choose to interact with swimmers. At these times, I watched in awe as these mammals from the sea took flight, their bellies pink with excitement as they performed acrobatic tail-over-head flips and spins of as many as eight revolutions with only three flicks of their powerful flukes.

"The theodolite is ready," announced Kristen, one of the research assistants, bringing me back to my present task.

"Okay, what've we got?" I questioned, referring to the number of humans surrounding the pod.

"One, two, three . . . twelve swimmers, five kayaks, and one Zodiac [a small motorized boat with rubber pontoons and a hard hull]," the spotter Matt replied confidently, while the data collector Kira quickly wrote the list.

Next Kristen called out the coordinates. "A group of three kayaks at ninety-one point three eight, two sixty-five point five four."

"Two sixty-five point five five," Kira repeated, checking her numbers.

"Five four," Kristen corrected. We continued to record the bearings on the circle of humans. I prepared myself for our first fifteen-minute data-gathering session—a period of intense concentration and confusion as we struggled to record all behaviors and movements of dolphins, people, and boats.

This amount of commotion around the dolphins is common at this time of the day. At 9:00 or 10:00 A.M. the kayaks begin to arrive, usually carrying curious tourists, groups of dolphin-swim tours, or workshop participants. From the cliff, the bay often resembles a three-ring circus with the main act in the center. This is the most difficult, most frustrating, and maybe the most important time to do research because of the level of activity. A typical scene includes fluorescent yellow and green and red and blue kayaks forming a ring, while swimmers in brightly colored suits, snorkels, and fins pursue slate-gray dolphins as they swim in and out of the circle. Swimmers snorkel face down looking at dolphins below, sometimes chase them, or bob like a cork watching where the dolphins have gone. When I am swimming with dolphins, this is the time I get out of the water. The feeling in the water changes from one of communion with fellow creatures to one of competition for attention.

Unfortunately, it is often this time of day, during the height of pandemonium, when the dolphins descend to rest, or at least try. They move closer together, breathe and move in unison, leaps and spins diminish, and they begin to "carpet" the sea floor, moving slowly in graceful sweeps above the sandy floor of the bay. It could be their most critical time of day. Think about it this way: it is like coming home from a fifteen-hour day to find people in your bedroom ready for a party. They insist on being entertained and will not leave for another three to four hours, which allows you only about six hours to rest or sleep.

Another sign that the dolphins are trying to rest is a behavior called "skirting," which becomes obvious when watched from above. Normally, people form a circle around the pod. When the dolphins are playing with us, they swim either within or in and out of the circle of people. As they seem to tire of this play and begin their descent into rest, they simply form a more cohesive group and swim around the perimeter of the area or "skirt" the human pod. It may be their way of saying "Goodnight," like polite hosts hinting that the party is over, rather than saying "Get out! We want to go to bed now."

As we began collecting information, I realized this was going to be a busy Saturday. And, as I feared, it was a frustrating morning to conduct any research. Keeping track of where the dolphins were and how many people were around them at a given moment was a challenge. Swimmers were coming and going, people jumped in and out of kayaks, the Zodiac tried to follow dolphins and avoid hitting kayaks and swimmers, who in turn moved around nervously trying to avoid being hit by the Zodiac. All this disruption seemed to be difficult for the dolphins as well. They tried several times to rest, skirting the group of people and slowly moving away. Each time they were startled from their rest by the noisy splashing and flailing about of humans and returned to being active at the surface with spins, leaps, and slaps. Finally, the bay began to empty, and it looked as if the dolphins might find some peace. But, at that moment, a bright yellow kayak separated from the human pod and chased the dolphins as they swam toward the far cove—the area where their privacy is normally ensured.

"Where are they going? They're chasing the dolphins," Kristen cried with a worried voice as she adjusted the instrument to take a reading on the offending kayak.

"Don't they know the dolphins are trying to rest?" asked Matt with a protective tone in his voice. I was so angry, I wished for a bullhorn at that moment to scream, *"STOP! And get away from the dolphins!"*

From 11:00 A.M. to 1:00 P.M., kayakers often seem to be the ones who present the greatest threat of disturbance for the dolphins as they move away to rest. Maybe it is because of the kayakers' middle perspective that they usually seem to be the last to leave the party. They are not in the water, so they can see where the dolphins go, as opposed to the swimmer who just knows the dolphins are gone. But the kayakers are so close they do not notice the pattern of skirting as we do on the cliff and thus do not take the hint. They often pursue dolphins all the way across the bay to a cove that seems to be the last bastion of privacy for the dolphins. Most swimmers will not swim the distance—about a mile—and the dolphins seem to know this from experience. However, it is within easy reach for a kayak. Not all people in kayaks behave this way and many are very considerate. It may be that people who have no previous experience with dolphins and therefore do not understand dolphin behavior are the ones most frequently using kayaks at this time.

We watched helplessly, collecting numbers and behaviors, as people in the kayak followed the dolphins to each place they surfaced, abandoned the kayak, and jumped in—effectively crashing through the dolphins' bedroom ceiling. Our only tool to help the dolphins was the information we collected. All we could do was hope we discovered something that could be used to help regulate the human activity in the bay.

"Fifteen minutes is up," the data collector informed us. We all relaxed, sitting back in our chairs and placing the equipment we had been using in our laps, and took a much-deserved break. I continued to watch as the insensitive kayakers remained unrelenting in their pursuit. I felt ashamed to be a member of the human race.

The tiny area of shade was quickly disappearing, and exposed skin on my left side burned in the afternoon sun. A breeze cooled my face and I listened to the rustling of dry grasses. As I relaxed in my chair during the fifteen-minute respite, I considered the kayak problem. My decision had been not to kayak if I were planning to swim near enough to the dolphins to glimpse their lives underwater. I wanted to meet the dolphins on their own terms in their environment. With dolphins, this means in the water and swimming. However, not everyone held the same opinion, and many days I watched dolphins being followed or surrounded by kayaks. I was beginning to know these dolphins. I felt a certain sense of responsibility to help protect them and to help them find time and space to rest. I felt frustrated and angry that not everyone behaved as if they were concerned about the dolphins' well-being. I sighed and wondered if we would ever learn to be proper guests in the dolphins' resting bay. Or maybe, I realized, it is enough to recognize that we are guests.

After our break was over, we switched jobs and waited for another interspecies group of humans and dolphins to form. I began recording data as the team called out dolphin behaviors and interesting comments about a woman wearing a white swimming cap and webbed gloves. "Wow, look at the dolphins right beside that lady!" cried Kira.

"Dolphins up," Kristen and Matt called in unison as the pod surfaced to breathe.

"They're still right with that woman! Now they're down," Matt reported. "Flippers up!" he joked, as the woman dove down with the dolphins, lifting her blue swim fins above the surface. When the dolphins resurfaced, they had moved away from the woman.

"They're going back to get her!" Kristen observed as the dolphins turned right around, swam back to the woman, and slowed down, milling around her as she swam next to them.

"I wonder what it is they like about her?" Kira asked.

I wondered myself. I remembered a few close encounters I had had while swimming with spinners, but never anything that lasted so long. I felt a pang of envy as our fifteen-minute observation period ended and the dolphins were still mingling with the swimmer in the white cap—coming up ahead of her and returning to include her in their pod.

"I wish I was down there right now!" one of the spotters said. I knew we all shared that sentiment as we sat, looking down at cool water and people swimming with dolphins. Although the research on the cliff was often uncomfortable and tedious, and I would much rather have been in the cool water observing dolphins, I am grateful for the unique perspective this elevated point of view afforded me. I learned to recognize the rhythms and challenges of dolphin life from a more sensitive, respectful distance.

As the noon lunch hour approached, the afternoon sun blazed in the sky, and the bay lay empty except for an unoccupied, fluorescent green kayak floating in the middle of the bay. Its owners floated face down about twenty yards away. From their position, I guessed they were watching dolphins swimming slowly along the sandy floor. Sure enough, a few minutes later the dolphins surfaced close to the snorkelers. I imagined hearing the series of soft whooshes as the dolphins breathed together as a pod. At long last, the dolphins could rest.

It is relatively easy to recognize a resting pod by watching its patterns of movement from the hill. During this time, the pod seems to act with a group mind. Resting dolphins remind me of a flock of birds all changing direction in the same moment. Most of the pod remains below the surface for a longer period of time, approximately five minutes, before returning to the surface to breathe together.

When dolphins rest, they travel in silence. It is believed that during this period they shut down their echolocation or natural sonar, move closer together—within touching distance—forming a "carpet" of dolphins along the bottom, and close one eye because that half of their brain is inactive.

After our second break of the day, we spent the remainder of our time on the cliff collecting much-needed baseline data—information on the dolphins in the absence of swimmers, kayakers, or motorboats. It was wonderful to watch the dolphins going about their lives without any interference from people. They seemed totally at peace in an environment that is foreign to us. I wondered what it would be like to be a member of such a close community, where every decision seems to be made for the good of every member equally and the pod as a whole.

At the end of our session, we packed up our equipment and prepared to climb down the cliff for the last time. I hoped the dolphins would have a peaceful afternoon. The late afternoon is typically the most peaceful time in the bay. Occasionally, a Zodiac full of tourists will come into the bay or a motorboat will pull a water skier near the pod, startling the dolphins out of their quiet slumber. Or one or two swimmers may approach the pod to watch from the surface as the dolphins swim slowly below. However, usually at this time of day the spinners are allowed to rest peacefully.

That afternoon, when we returned to the research house, we relaxed on the lawn overlooking the bay and watched for resting dolphins. Around 4:00 P.M., the dolphins began to stir from their rest. A few members of the pod started leaping and spinning as they became more excited. When they headed out to deep water for a night's feeding, I watched with delight as they performed spins, one after another, for miles as they left the bay. Silently, I wished them luck for a large feast that night.

During the two seasons I studied these spinner dolphins on the Big Island, I also experienced the dolphins in the water whenever I could. The difference in perspective between watching dolphins from the cliff and then swimming with them underwater was like the difference between flying over an old growth forest in a noisy, twin engine airplane and then hiking through the same forest, listening to its creatures and smelling the scent of the earth. And while I understood the potential for my behavior to disturb the dolphins, swimming with dolphins in the wild deepened my understanding of what I was seeing from the hill.

I learned, for example, that when the dolphins are milling within a group of people they are interacting closely with the humans. I could not have known this had I not had the underwater experience. I also could not have seen from my hill perspective the subtle ways in which dolphins may choose to initiate contact with swimmers. With their additional sense of echolocation, dolphins can locate a swimmer several hundred yards away. If they choose to interact or if they simply do not mind the presence of a swimmer, dolphins seem to allow themselves to be observed underwater. However, if they do not wish to be disturbed and swimmers cannot pursue them in a kayak or motorboat, the dolphins may simply remain outside the range of vision, engaged in a potentially exhausting game of chase. The swimmer is forced to play hide and seek with an unwilling participant, with the sound of dolphin chatter the only clue that the dolphins are still in the bay. Even after learning about the dolphins' "choice," I wondered if it were disturbing to the dolphins to have even one or two humans—the guests at the party—in the water during their resting period.

On one morning, when we were not on the hill collecting data, I slipped in for a swim. I had been thinking about the effects of humans on dolphins. These are perilous times for dolphins. Each time they leave the bay, the spinners face depleted food sources, debris, and pollution—both toxic and noise pollution—and fishing nets. Knowing all this, I still wanted something from them. I wanted to understand how it feels to be a dolphin, living in the sea. But I wanted to be extremely cautious and conscious of their well-being and not burden them with my human desires in the one place where they could rest peacefully.

I swam through the warm, clear water out to the middle of the bay. Filtered sunlight danced in rainbows on the rippled patterns in the sand below me. Dolphin sounds bounced around me as I traveled. Would I interpret their actions correctly, and was this an appropriate time to be swimming with them? I found myself apologizing again for being human, a member of a species who had brought them so many problems. I considered swimming back to the beach. And then, I saw them.

I do not remember how many were there, I only remember the mother and calf—two beautiful, tri-colored dolphins. The mother was slender in body with a slate-gray cape coloring her top from the tip of her long, narrow snout to her tail. A pale gray stripe graced her side from forehead to flukes, and her white belly gleamed. Across her eyes, a slate-gray mask colored her face from flipper to flipper. And she wore a gentle, enigmatic smile, like a Madonna of the sea. The calf, a miniature version of its mother, was still a little pudgy from baby fat. When she allowed the calf to swim between us, I felt certain the mother was not threatened by me.

As my distress shifted to wonder, the mother slowly moved closer to me, calf in tow. The rest of the pod had moved away and I was alone with the pair. She looked me in the eye and began to circle me, her calf still between us. I moved slowly in the water, careful not to startle her, and turned to swim on my side to watch her. *You have a beautiful baby,* I said silently, suppressing a smile to avoid filling my mask with seawater. Was she showing her calf to me, or was she showing me to him? I felt the tension that comes when it seems as if the heart can hold no more. I blinked away tears filling my eyes.

I reminded myself to breathe. With deep inhalations, my body calmed and I floated, suspended in sea, alongside the dolphins. My mind grew quiet and ceased its persistent thought, as if in meditation. Then, I heard a voice in my heart, *"Don't feel guilty for being human; you are the guardians."* I sensed the warm ocean swells sweeping over my back. The mother continued her steady gaze, while I remained transfixed by her gentle, knowing smile. It was as if she had every confidence in the world that I would do the right thing—as a steward. And in some primal space within, I allowed myself to accept my humanness and my place in the world.

Mother and calf circled me once more, and we swam along side by side while I watched the chubby calf fall into traveling position just behind mother's flipper. Then, as quickly as she had initiated contact, our encounter was over. I watched the pair vanish into the gray-blue mist of the bay. Swimming slowly back to shore through the clear, salt water, I listened to distant dolphin chatter.

Since that magical interaction, I have often wondered about the message *"You are the guardians."* I have considered its meaning for myself and for other humans. I have realized that, as a scientist studying dolphins, I am their representative. I study them to help protect them. As a steward, I feel a responsibility to share what I have learned about their lives and culture. As the dominant species—the guardians of the planet—we humans take on an extra burden when we choose to interact with and establish a bond of trust with another species. They may not understand our ways and may not be able to discern which of us offer friendship and which ones may bring them harm. The reverse is true as well. Whales and dolphins are large, wild animals, and they have been known to defend themselves when threatened.

However, through these encounters, we can learn from the creatures with whom we share the earth. Perhaps this bay, where people regularly interact with these remarkable creatures, could become a model of interspecies diplomacy—blurring the lines of who is taking care of whom. As students of these wise beings, whether we study them from a cliff or from the water, we can learn lessons about living in community, the importance of play, and how to care for and respect other species. I have to trust—like the mother dolphin—that humans will do the right thing and begin to consider the dolphins' well-being in decisions about how to share the oceans.

At this point in our evolution as a species, we stand on the path at the beginning of an uphill climb, which is indeed treacherous. Respecting wild dolphins could be one small step toward our deeper evolution. If we can learn to respect them, realizing that they exist for themselves, not to benefit us, and if we can learn to interact with them as equals, our worldview may begin to shift. If we can learn to listen, we may be able to save ourselves as well as other species.

ROCHELLE CONSTANTINE
AND SUZANNE YIN

Swimming with Dolphins
in New Zealand

New Zealand is one of the few places in the world where people are not merely permitted to swim with wild dolphins but are given ample opportunity to do so as part of an expanding tourism industry. In 1989, the first commercial tour operator started taking paying passengers to see and occasionally swim with dusky dolphins in the waters off Kaikoura. Today people can swim with a variety of species in approximately twenty-five places around coastal New Zealand.

In the early years of dolphin tourism, the New Zealand government theorized that permitting companies to take people out to interact with marine mammals in their natural environment would enhance human understanding of these fascinating animals and thereby result in greater sensitivity and conservation efforts. The government further decided that a permit-based system would create some control over the number of people interacting with marine mammals on a regular basis and thereby provide some protection from overexploitation.

Subsequently, New Zealand has become one of the most popular places in the world for some people to fulfill their lifelong dream: to swim with wild dolphins. Tens of thousands of people now attempt to swim with dolphins every year in New Zealand. Unfortunately, in our desire to get close to these animals, we have neglected the most fundamental question—do dolphins want to swim with us?

Our combined research adds up to twelve years spent trying to unravel the complexities of dolphin-based tourism. We have spent countless hours standing on a hill or driving a small inflatable boat collecting data on the movements and behavior of dolphins, swimmers, and boats. It is only recently that we have been able to come to some conclusions about what it may mean to the dolphins to have boats and swimmers coming to interact with them on a regular basis.

One of the most interesting aspects of our research is that the three different species of dolphins we have studied (bottlenose, dusky, and common dolphins) often show markedly different responses to boats and swimmers. It appears that certain types of boat handling that cause severe avoidance responses in one species will result in the most minor response by another. This is not surprising, considering how variable dolphin behavior is. Dolphins live in groups ranging in size from a few individuals to several hundred; they forage collectively or individually; and they live in the open ocean vulnerable to predation by sharks and killer whales or in shallow waters vulnerable to entanglement in fishing nets. For researchers trying to understand the effects of human-dolphin interactions, the extreme variability in dolphin society alone makes this a complex field of study.

We conduct our ongoing research in two very different parts of New Zealand. Because we study different species, we use different research methods. During her research on dusky dolphins, Suzanne remains primarily land-based. In Kaikoura, the dolphins come close to shore into a large open bay, which allows perfect viewing of the large groups of duskies from land.[1] Viewing dolphins from shore means that dolphin behavior can be tracked while preventing potentially harmful impact. This land-based approach allows Suzanne to track changes in the speed and direction of boats and dolphins, and to determine how boats and swimmers affect the dolphins.

In contrast, the bottlenose dolphins in the Bay of Islands weave their way throughout the 144 islands, allowing no place where land-based observation can be made for Rochelle's study. Research of the bottlenose dolphins requires closer study, which allows more detailed analysis of the effect of swims on individual dolphins. Although the boat is a potential disturbance to the dolphins, Rochelle has found that careful boat handling can habituate the dolphins to the presence of the 4.7-meter research vessel, and the dolphins will engage in behaviors such as resting that they generally will not do when other boats or swimmers are present.

BOTTLENOSE DOLPHINS
IN THE BAY OF ISLANDS

Since early 1994, Rochelle's work has involved spending a lot of time with a population of approximately 450 bottlenose dolphins ranging along coastal Northland, New Zealand. New Zealand has some of the largest coastal bottlenose dolphins in the world, at approximately three to three and a half meters in length. Her research is focused on the population size, habitat use, and ranging patterns of this population as well as evaluating the effects of tourism on them. Rochelle has observed hundreds of swim attempts and watched the dolphins' behavior in the presence and absence of commercial swim-with-the-dolphin and recreational vessels. All these hours have been used to explore the question of how the swim-with-the-dolphin industry affects the dolphins' lives.

The majority of people who go swimming with dolphins in the Bay of Islands join a permitted tour where the equipment, expertise, and knowledge on how best to swim are supplied. But, even under these protected conditions, research has shown that the impact of swimming with dolphins is much more complex than might be suspected. The dolphins respond extremely differently depending on where the swimmers are placed.[2] For example, if swimmers enter the water in the dolphins' path of travel, no matter how slow the dolphins are moving they will almost always change direction and avoid swimmers. For common dolphins this happens 86 percent of the time, and for bottlenose dolphins 77 percent of the time.

However, if swimmers enter the water off to the side, dolphins can choose to come near the swimmers or not. The avoidance rate in this case is only 21 percent for common dolphins and 2 percent for bottlenose dolphins. The risk for the humans in giving the dolphins a choice is that the dolphins do not always choose to interact, no matter how much we want them to come near. In fact, when given the choice, bottlenose dolphins will approach swimmers 34 percent of the time; the rest of the time they will continue to engage in the activity prior to swimmers' entering the water. When swimmers entered the water off to the side of common dolphins, the dolphins never approached the swimmers. This method of swimmer placement is the true test of whether dolphins want to interact with humans or not, and it appears that, given the choice, the majority of the time they would rather go about their normal lives.

Perhaps the most interesting finding to date is that, over time, the bottlenose dolphins in the Bay of Islands are avoiding swimmers more and interacting less. Looking at possible explanations for this change in behavior, we conducted a photo-identification study of individuals. This study has shown that, although the dolphins are not resident in the Bay of Islands, they return there throughout the year. Over time this ranging behavior has resulted in individuals being exposed to multiple encounters with commercial swim-with-the-dolphin tours, and it appears that swimmer placements that do not give the dolphins a choice have led to significant avoidance.

On those occasions when dolphins do interact with humans, it appears that the more comfortable and competent someone is in the water, the greater response they are likely to get from the dolphins. On occasion dolphins seem willing to include us in their play behavior, but we should not assume that they are there to entertain us. Unfortunately, we have also observed some interactions between humans and dolphins that have resulted in harassment. People leap off the side of their boat directly on top of dolphins, lunge for them as they swim past, and drive at high-speed in circles in an attempt to get them to jump. Many times people do not realize that their misguided efforts to interact with the dolphins are either dangerous or disturbing to the animals.

All of the examples described above result in the dolphins changing their behavior and immediately leaving the area. Fortunately, education in the Bay of Islands has resulted in an improvement in behavior around dolphins. The commercial operators, who spend by far the greatest amount of time with the dolphins, constantly modify their behavior to minimize their impact on the dolphins. Their livelihood depends on this sensitivity: they must pay attention to the dolphins' behavior if they want to keep the dolphins nearby.

DUSKY DOLPHINS IN KAIKOURA

Suzanne has conducted her fieldwork near Kaikoura, a small town located on the east coast of the South Island, about 1,100 kilometers south of Rochelle's study site. In this area, Suzanne has studied the movement patterns and acoustic behavior of dusky dolphins, commonly called duskies.

They are only found in the Southern Hemisphere, around South America and South Africa, as well as off New Zealand.[3] Smaller than the bottlenose dolphins that Rochelle studies, adult dusky dolphins are usually less than two meters in length and are often found in larger groups than bottlenose dolphins—sometimes in groups of over 1,000 animals.

Kaikoura is probably one of the best places in the world to see dusky dolphins, sperm whales, and many species of seabirds. It may also be a crucial environment for duskies to spend their summer months. Kaikoura is one of the few areas around the New Zealand coastline where deep, nutrient-rich water is found close to shore. The bays off the Kaikoura Peninsula provide calm, protected waters for dusky dolphins to rest and socialize during the day. But the period when duskies rest is precisely when people would like to interact with the dolphins. What is the effect of the tourist vessels and swimmers on the behavior of the duskies?

Most of our work has entailed watching dolphins from shore, tracking their movements and those of tourist and research vessels and swimmers by use of a theodolite. Although the research is conducted much farther away from the duskies than Rochelle is from the bottlenose dolphins she studies, we are able to observe dolphin behavior before, during, and after the arrival and departure of the boats. This means that we can see whether the dolphins have changed their behavior when boats and swimmers are near, but we keep from disturbing the dolphins by our presence.

The research in this area focused on the movement patterns of small groups of fewer than twenty-five dolphins and their behaviors in relation to the presence or absence of boats. Choosing distances of 100 meters and 300 meters, critical distances in New Zealand's Marine Mammals Protection Regulations (1992), the research found that small groups of dolphins swam in much straighter lines when boats were nearby or had just departed than before the arrival of the boats.[4] Dolphins also made small, erratic changes in their direction of travel when boats were within 300 meters. At this point we cannot conclude that these short-term behavioral changes are detrimental to the dolphins' long-term health. We will have to continue our work to determine whether there are any long-term changes in distribution and behavior of these animals. The rest of the research focused on describing the types of sounds that duskies make. After locating a group of duskies, we would lower an underwater microphone, known as a hydrophone, into the water. Dusky vocalizations—squeaks, squawks, and whistles—resemble the background noise at a dinner party, with one type of sound, burst pulses, dominating much of their repertoire. Whistles are the most infrequent vocalization type. However, there seems to be a tendency for duskies to whistle more when swimmers are in the water than at other times. Duskies will whistle at a rate of 0.62 whistles per minute when recorded in the presence of swimmers, versus only 0.01 whistles per minute during all other times (that comes to only twenty whistles in over twenty-five hours of recording). We do not yet know why duskies whistle more when swimmers are in the water, but this is one behavior we will investigate more closely in the future. It is possible that duskies are separated by the presence of the boats and swimmers, and so use the whistles as contact calls between dolphins. It is also possible that there is a higher level of activity for the duskies when swimmers are in the water.

Dolphins live in an acoustic world frequently visited by swim-with-the-dolphin boats. The dolphins must rest, socialize, and forage—while we maneuver our boats around them to find the "best" position to get into the water. When we do enter the water, it is often directly in front of them and typically involves a lot of noise and splashing as we try to "entertain" them and coerce them into interacting with us. Taking tours out to see these fascinating animals is a great way to educate people about the marine environment, *but not if it is at the expense of the dolphins leading normal daily lives.* Studies like ours are beginning to unravel some of the questions about whether the dolphins actually want to swim with us. With the results to date, it appears that not all dolphins are interested, that they appreciate the space to choose whether or not to come over to swimmers, and that they appreciate periods when there are no boats with them. These may sound like commonsense results, but unfortunately in our desire to get closer to dolphins we sometimes neglect the most important thing of all: respect for these fascinating creatures.

NOTES

1. B. Würsig, F. Cipriano, and M. Würsig, "Dolphin Movement Patterns: Information from Radio and Theodolite Studies," in *Dolphin Societies: Discoveries and Puzzles*, ed. K. Pryor and K. S. Norris (Berkeley: University of California Press, 1991).

2. R. Constantine, "Increased Avoidance of Swimmers by Wild Bottlenose Dolphins (*Tursiops truncatus*) Due to Long-term Exposure to Swim-with-Dolphin Tourism," *Marine Mammal Science* 17, no. 4 (2001): 689–702.

3. R. L. Brownell and F. Cipriano, "Dusky Dolphin," in *Handbook of Marine Mammals*, vol. 6, ed. S. H. Ridgway and R. Harrison (San Diego, Calif.: Academic Press, 1999).

4. S. E. Yin, "Movement Patterns, Behavior, and Whistle Sounds of Dolphin Groups off Kaikoura, New Zealand" (Master's thesis, Texas A&M University, 1999).

Troubling Tursiops: Living in Harmony with Kindred Spirits

A few years ago I was standing in line at a grocery store when I heard a young girl tell her friend that she had just gone swimming with dolphins when on holiday in Hawaii. She had a great time, but there was a slight pause when her friend asked her about what the dolphins might have felt about all of this. Did they enjoy her touching them or riding on their back? Did they really like being bothered? I smiled and walked away, thinking, *Now isn't that what we want, young kids asking questions about ethics even if they don't know that's what they're doing.*

This encounter was among the reasons I decided to write a children's book on animal ethics, titled *Strolling with Our Kin*, and edit a book on animal feelings, called *The Smile of a Dolphin: Remarkable Accounts of Animal Emotions*. I also ask these sorts of questions—what would it be like to be a certain animal, what do they feel, and what would they have to say about what humans do to them—when I work with local elementary schools as the regional coordinator for my good friend Jane Goodall's Roots and Shoots program. Maybe we adults also need to do a bit more soul-searching about our never-ending intrusions into the lives of nonhuman animal beings (hereafter animals). So, for example, should humans swim with dolphins? Even if it is not in the dolphin's best interests, is it okay to swim with them if it helps yourself or a family member? A friend? A stranger? Should we ever intrude on other animals if it did not have some positive effect on them? These nagging questions rapidly lead us into difficult terrain, the negotiation of which requires open minds and especially open hearts.

Writing about animal ethics is bound to raise hackles and bring one's critics out of the woodwork, but this is just the price of doing business. Many people are deeply passionate about the imbalance of human-animal interactions—we get a lot and they get very little. Although some of my views are considered "radical," especially by some of my ivory-tower colleagues, I wonder why it is considered radical to argue that we should let other animals be, to allow them to have their own lives as much as is possible in this human-dominated and anthropocentric world. I am confused, because I am not trying to put my colleagues out of business but, rather, asking them to think more deeply about how we trouble other animals over and over again. I have really had enough. Thus, I am willing to open myself to criticism, to be vulnerable for expressing views that are not part of mainstream science.

In *Strolling with Our Kin*, I laid out some questions that force people to confront head-on some very difficult questions. These included: Should we interfere in animals' lives when we have spoiled their habitats or when they are sick, provide food when there is not enough to go around, stop predators in their tracks, or translocate individuals from one place to another, including zoos and aquariums? Should our interests trump theirs? Is it permissible to play music to dolphins as long as they can move away? Is it permissible to swim with dolphins if we can determine that the dolphins enjoy it? Is it permissible to swim with dolphins if only the humans benefit? Is it allowable to use motor boats and water ski where they live? Should we interfere with serious aggressive encounters between two dolphins?

The list goes on endlessly. The question of when humans *should* intrude is a difficult one. However, just because we *can* do something does not mean that we *should* or *have to* do it. Furthermore, just because some intrusions may be *relatively* less benign than others, this sort of claim places us on a very slippery slope and can lead to thoroughly selfish anthropocentric claims. Even in situations when we have good intentions, good intentions are not always enough.

"Troubling tursiops" has two meanings. The first twist on the word "troubling" concerns how humans trouble *Tursiops* (the genus for bottlenose dolphins, the most sought-after dolphins for captive display and swim programs), how we intrude into their worlds, how we bother them (even unintentionally). The second concerns the troubling and complex issues that arise when we ponder what sort of intrusions, if any, are permissible.

Here I want to write about some troubling issues that underlie the use and exploitation of dolphins (and other animals) and ask some general questions. I will not answer many of the questions definitively but, rather, will try to provide some guidelines for proactive decision-making; all too often we are trying to rectify messes that we have created, and proactivity needs to become the modus operandi for future actions. For many questions about how animals should be treated by humans, there are not any "right" or "wrong" answers. However, there are better and worse answers. Open discussion will help us make better decisions on behalf of all animals.

It is a given that humans are everywhere. Give us an inch and we take a foot and then some. There are not any places on earth that are not influenced by human activities. Yet we are part of nature, whether we like it or not. We are an integral piece with awesome responsibilities that can no longer be pushed aside for convenience, or because someone else will clean up the messes we leave.

When many humans interact with nature, they usually wind up redecorating it, selfishly. Intentionally or not, it is as if humans have an inborn urge to reshape nature, to expand their horizons. We just cannot stop ourselves, and little else does, even the blatant results of our trying to dominate—manage, control—our surroundings. We move animals around as we move furniture, and we redecorate landscapes with little concern for maintaining biological integrity. Even during strolls in pristine forests, swims in oceans, or forays in the sky, most humans are detached and alienated from the majesty of their surroundings. They do not seem to love nature deeply.

Because humans have incredible power to dominate animals, our animal kin depend on our goodwill and mercy. Animals depend on humans to have their best interests in mind. *We can choose to be intrusive, abusive, or compassionate.* We do not have to do something because someone else wants us to do it "in the name of science." We do not have to do something just because we can do it. Each of us is responsible for our choices. Science is not value-free, and we all come to our work with an agenda.

Some of my own guiding principles for nonconsumptive interactions with animals include: (1) putting respect, compassion, and admiration for other animals first and foremost; (2) taking seriously the animals' points of view; (3) erring on the animals' side when uncertain about their feeling pain or suffering; (4) recognizing that almost all of the methods used to study animals, even in the field, are intrusions on their lives—that much research is fundamentally exploitive; (5) recognizing how misguided speciesism is, especially concerning vague notions such as intelligence and cognitive or mental complexity for informing assessments of well-being; (6) focusing on the importance of individuals; (7) appreciating individual variation and the diversity of the lives of different individuals in the worlds within which they live; (8) appealing to what some call questionable practices that have no place in the conduct of science, such as the use of common sense and empathy; and (9) using broadly based rules of loyalty, trust, and nonintervention into animals' lives.

My view of human-animal relations makes me uneasy about claiming that any group of animals is special. Unique yes, special no. I do not think that dolphins or whales are any more special than mice, robins, goldfish, alligators, shrimp, or ants. Being sentient, capable of experiencing pain and pleasure, does not demand that we value one life over another. Nor is being cute or warm or furry sufficient reason for special consideration. Rather, I have come to believe and feel deeply in my heart that all life is valuable and that all life should be revered. *Animals are not resources or property with which we can do what we please; their lives matter very much, and they should be firmly entrenched in our moral community.*

Nonetheless, when I come back to solid ground I realize that humans are a curious lot, and our intrusions into the lives of other animals, whether intentional or inadvertent, have significant impacts on their well-being. And there is much information that shows the effects (often unintentional) we have on animals as we conduct research or engage in recreational activities, including photography, hiking, and traveling (eco-touring). Many of these findings apply to other situations and species, including dolphins.

We do not have to handle other animals to influence their lives. "Just being there" can have large and long-term effects. For example, elk and numerous other animals avoid skiers and thus change their patterns of travel and foraging. Research performed by my students and myself has shown that humans have a large influence on prairie dogs: individuals who have lots of contact with humans are less wary of their presence than individuals who do not. The same is true for various deer. Magpies, common local birds not habituated to human presence, spend so much time avoiding humans that this takes time away from essential activities such as feeding. Researchers interested in feeding patterns must be sure that their presence does not alter species-typical behavior, the very information they want to collect.

Many people adore young animals and try to get close to nests or dens without disturbing residents. However, nests visited regularly by humans can suffer higher predation than nests visited infrequently. This has been shown in various birds, including ducks. Yes, ducklings are cute, but getting too close can be lethal. It seems that birds accustomed to intrusions by humans who do not kill them subsequently allow other animals, including natural predators, to get too close.

People also enjoy watching animals from cars, boats, or airplanes. However, the noise and presence of vehicles can produce changes in movement patterns (elk), foraging (mountain sheep), and incubation. In swans, the noise and presence of cars result in increases in the mortality of eggs and hatchlings. Once again, these effects are not obvious when they occur, but data show they are real.

Often, researchers want to know who's who, so they place colored tags on the legs of birds or fit mammals with ear-tags or radio-collars that transmit a signal unique to each individual. These techniques can have a major effect on the behavior of the adorned animals. Placing a wing-tag on ruddy ducks leads to decreased rates of courtship and more time sleeping and preening. Thus, local populations may decline. Brightly colored radio-collars on sheep lead to increased predation.

Now what about dolphins? People often justify their interactions with other animals by arguing that the animals get as much or more out of interacting with humans than do the humans. Indeed, this line of argument can lead to many different conclusions, depending on the humans' point of view. Perhaps this reasoning underlies the absurd claim that by keeping dolphins in captivity we are providing them with free meals and health insurance, and their lives are better than those of wild individuals. Indeed, this mentality led to the exploitation of Flipper and the development of numerous big and very profitable businesses—"dolphinaria"—each with its obligatory Flipper.

Most descriptions of dolphins and other cetaceans portray them as highly intelligent and sentient animals with remarkable social and cognitive skills. They seem to empathize with other individuals and mourn the death of their friends. Why then do we feel comfortable intruding into their worlds? Would we feel the same if we thought that they were stupid and unemotional?

I am thinking once again about the two schoolgirls who I met in the grocery store, one of whom had recently swum with dolphins. What do you think about "swim-with-the-dolphin" programs and the many touchy issues that are raised by these activities? Such reasons as "it's fun" or "the animals are cute" do not bear on the complex issues at hand.

I cannot possibly cover them all here, but let me consider a few difficult areas of the debate. First, a good deal of attention has been given to the question of whether or not human encounters with dolphins might have a negative effect on the dolphins—both in captivity and in the wild. What about dolphin-assisted-therapy (DAT) programs? DAT programs raise many sensitive issues because of their potential to help humans in need. Nonetheless, while some researchers claim that DAT is an effective therapeutic intervention for several disorders (e.g., autism, cerebral palsy, mental disability), others disagree, and the jury is still out. Alan Beck and Aaron Katcher concluded, in their book *Between Pets and People*, that "At present we have no evidence that contact with dolphins has any advantage over contact with any other kind of animal for any purpose. Having said that, it must be recognized that the cult of the dolphin is so firmly established in our culture that no evidence will dissuade people from believing that swimming with these engaging animals is not the best means of coming into direct contact with the healing powers of Mother Nature."[1]

Another important question is whether or not programs that involve interactions with captive dolphins help to educate people about these wonderful animals. There does not seem to be any evidence that interactive captive programs with dolphins are more effective educationally than non-interactive programs. Indeed, these programs may send the message that it is acceptable to take animals from the wild and bring them into captivity and keep them in small tanks or oceanic pens or tanks where they are bored and deprived and needlessly die. These are horribly impoverished environments.

It is truly essential to know how our intrusions influence the lives of animals and to share this knowledge so that we do not inadvertently bother them. To disseminate information about the "human dimension," administrators of zoos, wildlife theme parks, aquariums, and areas where animals roam freely need to inform visitors how they may influence the behavior of animals they want to see. Tourism companies, nature clubs and societies, and schools need to do the same. By treading lightly, humans can enjoy the company of other animals without making them pay for our interest in their fascinating lives. Our curiosity about other animals need not harm them.

As a scientist I am often criticized for being anti-science. But it really is in the best traditions of science to ask questions about ethics. It is not anti-science to question what we do when we study other animals. Ethics can enrich our views of other animals in their own worlds and in our different worlds, and help us to see that variations among animals are worthy of respect, admiration, and appreciation. The study of ethics can also broaden the range of possible ways in which we interact with other animals without compromising their lives. Ethical discussion can help us to see alternatives to past actions that have disrespected other animals and, in the end, have not served us or other animals well. In this way, the study of ethics is enriching to other animals and to ourselves in that we may come to consider new possibilities for how we interact with other animals. If we think ethical considerations are stifling and create unnecessary hurdles over which we just jump in order to do what we want to get done, then we will lose rich opportunities to learn more about other animals and also ourselves.

Many scientists are also control freaks. They perform highly controlled experiments in fabricated situations and wonder why the solution of real-world problems eludes them. Many scientists feel uncomfortable when they cannot control variables, when unexplainable phenomena trip them up despite their conducting carefully controlled experiments. They want certainty; they want to be able to establish causal relationships even when it is clear that these sorts of exercises fail as often as they succeed for large-scale and complex problems such as those resulting from human interactions with, and our influences on, nature. The study of human relationships with nature provides as challenging a field as there is in terms of the complexity of the numerous connections among different variables. Cause-and-effect relationships are often difficult to establish, and it is necessary to accept the challenge that faces us and to revel in the mysteries and awe of nature.

Clearly, humans and many other animals are going to cross paths with cross purposes. And when this happens, nonhumans usually lose. Allowing human interests always to trump the interests of other animals is not the road to travel if we are ever to solve the numerous and complex problems at hand.

We need to learn as much as we can about the lives of wild animals. Being the same as us is not a reason to respect and value animals, and being different from us is not a reason to dismiss other animals' lives. Our ethical obligations to other animals require us to learn about the innumerable ways in which we influence their lives when we study them, move them from one place to another, or imprison them in steel cages and tanks of water, regardless of size. Although animals are unable to consent to or refuse our intrusions into their lives, it is useful to ask what they might say if they could do so. Careful observations can tell us a lot about what they are feeling and saying in a given situation. We just need to open our senses and hearts to them.

It is truly a privilege to study animals and to share their worlds and lives with them. As we learn more about how we influence other animals, we will be able to adopt proactive, rather than reactive, strategies. Part of learning entails changing our practices and asking, "Would we do what we did again, or have we learned something that can make other animals' lives better?"

I have used the term "deep ethology" to stress that people should recognize not only that they are an integral part of nature but also that they have unique responsibilities to nature. "Deep ethology" also means respecting all animals, appreciating all animals, showing compassion for all animals, and feeling for all animals from one's heart. "Deep ethology" means resisting speciesism. A deep respect for animals does not mean that just because animals are respected we can then do whatever we want to them.

A compassionate ethic of caring and sharing is needed now, not later when it is convenient. Sensitivity and humility are essential components of our guiding ethic. Expanding our circle of respect and understanding can help bring us all together. The community "out there" needs to become the community "in here"—in our hearts. We are other animals' guardians, and we owe them unconditional compassion, respect, support, and love. We may have control and dominion over other animals, but this does not mean that we have to exploit and dominate them. Even if laws and animal-care regulations permit certain activities, our behavior toward other animals should be motivated not by law but by morality, compassion, and love.

My views on animal use are indeed restrictive: we are not nature's keepers if this means that we "keep nature" by dominating other animals using a narrow anthropocentric agenda in which other animals are objectified—referred to by numbers and not by names, transformed into points on a graph—and their worldviews discounted.

So, where to go from here? Our starting point should be that we will not intrude on other animals' lives unless we can argue that we have a right to override this maxim, that our actions are in the best interests of the animals irrespective of our desires. When unsure about how we influence the lives of other animals, we should give them the benefit of the doubt and err on the side of the animals. It is better to be safe than sorry.

I have raised a number of important and difficult questions that I hope will compel readers to come to terms with the complex interrelationships among all animal life—indeed, all life—on this wondrous planet. We should not impose our ways on other animals for selfish, anthropocentric ends. We are the voices for all animals, and all life on earth will benefit from carefully listening, learning, and appreciating each and every individual's own world-view. We will all benefit greatly by expanding our horizons by including all animals in our hearts, by informing all of our actions with deep feelings of respect, compassion, and love. We all make choices, and we all are responsible for them.

We are very fortunate to share this planet with wonderful animal friends. We are all partners in a joint venture. We can no longer be at war with the rest of the world; no one can be an island in this intimately connected universe. Taking animals seriously will result in a deeper examination and understanding of their points of view, and this knowledge will be important for further studies of interactions between humans and other animals.

If we forget that humans and other animals are all part of the same world and are deeply interconnected at many levels of interaction, then when things go amiss in those interactions and animals are set apart from and, inevitably, below humans, I am sure we will miss the animals more than the animal survivors will miss us. The interconnectivity and spirit of the world will be lost forever, and these losses will make for a severely impoverished universe. A world without nonhuman animals would be a dreadful one.

The fragility of the natural order requires that we all work harmoniously so as not to destroy nature's wholeness, goodness, and generosity. All life interacts with and influences all life. We are not alone. The separation of "us" from "them" engenders a false dichotomy, the result of which is a distancing that erodes, rather than enriches, the possible numerous intimate relationships that can develop among all animal life.

Is there a solution for the situation we have created, is there hope for the future? Surely there is; holistic and heart-driven compassionate science needs to replace reductionist and impersonal environmental science. Creative proactive solutions drenched in deep caring, respect, and love for the universe have to be developed to deal with the broad range of challenging problems with which we are confronted. Respect for all animals must prevail. To this end, Jane Goodall and I formed the group Ethologists for the Ethical Treatment of Animals (www.ethologicalethics.org).

Love is an essential ingredient in the recipe for reconciliation. Its power must not be underestimated as we forge ahead to reconnect with nature. Passionate impatience and quick fixes must be tempered with compassion and love. Often we sit around and ponder the crises as wilderness and wildness disappear right in front of our eyes. California condors, gray wolves, Canadian lynx, and numerous but less appealing and less known animals have been brought to the brink of extinction by human arrogance and greed. Primary forests have all but disappeared. Shopping malls and parking lots take precedence over the lives of such threatened species as black-tailed prairie dogs, whose families are decimated by bulldozers drowning them in their underground homes.

It is easy to throw up our hands and give up hope. We have truly made some egregious errors. But we need to remain hopeful if there is any possible salvation, if we are ever to reconnect to nature as our ancestors were connected during their long evolutionary history. Indeed, life as we now know it is only a blip in time. Rather than take a doomsday view that the world will not even exist in 100 years if we fail to accept our unique responsibilities, it is more disturbing to imagine a world in which humans and other life coexist in the absence of any intimacy and interconnectedness. Surely we do not want to be remembered as the generation that killed nature. Nature is our unconditional friend, and reconnecting with nature can help overcome alienation and loneliness. I know I would miss other animals and natural environs more than I would miss a finger or two.

Solid science can be driven by one's heartstrings. Saturating science with spirit, compassion, and love will help bring science and nature together into a unified whole. In this way, magnificent nature—the cacophony of her deep and rich sensuality—will be respected, cherished, and loved. Reconnecting with nature can help overcome alienation and loneliness.

Cetaceans are not human, but this does not mean that they are not sentient, that they cannot experience excruciating pain and suffering, or that they do not have rich intellectual and deep emotional lives. Cetaceans are closely linked to the wholeness of many ecosystems, and how they fare is tightly associated with how communities and ecosystems fare. By paying close attention to what we do to them and why we do what we do where and when we do it, we can help maintain the health of individuals, species, populations, and ecosystems. There is no substitute for respecting animals, treating them with heartfelt compassion, and loving them for who they are in this magnificent and awe-inspiring world. Could anyone reasonably argue that a world with less cruelty and destruction and more compassion and love would not be a better place in which to live and to raise children? I don't think so.

It seems pretty simple to me. When animals lose, we all lose!

NOTE

1. Alan Beck and Aaron Katcher, *Between Pets and People: The Importance of Animal Companionship* (West Lafayette, Ind.: Purdue University Press, 1996), 140.

PART IV

"Necessary Kindness": Our Shared Future

Our future with dolphins is increasingly dependent upon our willingness to extend kindness toward all species as well as our own. Joana McIntyre Varawa, who compiled the cetacean classic *Mind in the Waters*, exemplifies such kindness in "Loving with an Open Hand," in which she imagines a future that calls for people to let dolphins live in peace.

Kathleen Dudzinski, an innovative researcher who studies dolphin communication, prompts us to reevaluate the ways in which we interact with dolphins. In "Letting Dolphins Speak: Are We Listening?" we learn that the most important prerequisite to solving the "riddle" of dolphin communication may be found by being "quiet" long enough to really see and hear them.

Perhaps the necessity of kindness is no greater than when seeking dolphins for healing. Horace Dobbs, who has explored the healing power of the dolphin-human bond through alternatives to actual dolphins, raises in "Dolphin Healing from Afar," the issue of the almost insatiable desire for human healing from a literally exhaustible number of dolphins.

In her essay "With a Little Help from My Friends," Cathy Englehart, a chiropractor, also offers a visionary alternative to our intrusion into dolphin societies and habitats. Instead, we may engage them with our imagination—without interrupting their daily lives.

Paul Spong, renowned co-founder of Greenpeace and orca researcher, and Helena Symonds, a pioneer of noninvasive orca research, metaphorically describe the possibilities of the dolphin-human bond in "The Ocean's Chalk Circle." They predict that the future of cetacean protection will depend on people's ability to expand their self-interest to include the well-being of other species.

In "Necessary Kindness," Brenda Peterson interviews Jean-Michel Cousteau. Jean-Michel's vital work for the oceans continues the impassioned work of his late father, Jacques Cousteau, to educate and conserve, because "people protect what they love."

Lastly, in "Advice to a Prophet," the respected former Poet Laureate of the United States Richard Wilbur asks a question that could be the epigraph for this book: "What should we be without / The dolphin's arc . . . ?"

Loving with an Open Hand

Every time I walk with my dogs, I am impressed with the brilliance of their attention. Their minds seem clear and present, and as I muddle from rock to rock with my multitude of confusions and questions, the dogs are untroubled and far more able than I to respond appropriately and quickly to whatever occurs around them. Just what is going on here? Most of the time I hardly even know what to eat. Will this food poison me, will it be good or bad for my health, what is hidden that I should know about? Something has happened to my human mind to make it murky and full of doubt. It is easy to blame it on the "culture" or "them," but it seems like the dogs are, in truth, more intelligent than me. Never mind that they cannot invent a nuclear bomb, build high-rise office buildings, or communicate via the Internet. Those accomplishments seem less and less promising and more and more like mistakes along the way. Piss points on the path of progress.

Now, I well know that the dogs are not dolphins, and the subject of this book is dolphins. But long-term relationships with dolphins are limited by our inability to maintain in the sea and by the fake understanding obtained by keeping dolphins captive. After all, learning from captive animals is much like learning about humanity by watching men in the exercise yard at San Quentin. There just is not enough information to go on.

Some 400 years ago, Francis Bacon, one of the fathers of modern science, delivered his now-famous advice that "the secrets of nature reveal themselves more readily under the vexations of art than when they go their own way." "Perfect!" declared the august members of the English and French Academies, "let's vex nature and see what she reveals." Well, we have been vexing nature for these many years, and she is definitely vexed. We have reached an epoch on the time line of human life where most of our distinct abilities seem to lead us further into disaster. We cannot manage even the simplest solutions but instead veer drunkenly from one massive answer to another, wreaking havoc on our path. Clearly, the other animals know something we have forgotten, or never have known; they know a way of living that, from this gloom of our lost understanding, seems to point a way to a lessening of the reverberations—the pendulum of change.

Early on I fantasized about dolphins; why not? They are so sleek and unknowable. It was easy to load onto them any random desire I might have for affection, for company, for a simpler connection with the angular protuberances of this life. Feeling confused and lonely in the presence of my fellow humans, and disappointed in what civilized society offered, I sought a connection in a different arena. I was looking for knowledge that came from a deeper source, a source I called Nature.

My first meetings with dolphins were inconclusive. They did not prove anything, were vague, and could barely be interpreted. I swam in a muddy lagoon at seaside (hardly the word for my frantic splashing and thrashing in the opaque water) with two captive dolphins, Florida and Liberty, in Key West, and having no aquatic skills, I worried about whether the dolphins would let me return to shore, for they were described as not wanting to let people out of the water once they got in to "play" with them. Burdened by this particular anxiety, I muddled around in the dark half terrified of the dolphins, and soon felt beside me a dolphin, who took me from behind my knee and gently towed me to the rubber boat that served as our way of getting in and out of the lagoon. I was totally mystified by what part of their anatomy the dolphin had used to take hold of me. Was it a flipper? Had he or she known I was worried? Just what had happened in that "encounter with a dolphin"?

A few months later, there was a grand media event involving myself, my friend Joana McClure, and the actress Candice Bergen, who was going to help us save whales. Candice, who I had never met, flew up from Los Angeles to San Francisco, where we were all going to "swim" with two dolphins at the Steinhardt Aquarium who were in a tank that was thirty feet across; hardly a venture demanding serious waterwomanship on our parts and an environment demanding the greatest restraint on the part of the dolphins. Candice had just met someone and was in love with him and did not want to get her hair wet, as she had a date that night. Joana, used to being part of a vast entourage (once having traveled in the Bob Dylan parade), relaxed and enjoyed basking out of the limelight, and I worried about drowning and the dolphins. I think I slipped into the side of the pool for less than a minute; the photographers got their pictures of Candice, and I can't remember what Joana did. The dolphins were an absolute blur; a prop.

Myth mixed grandly with experience. I pursued and studied myth because it seemed more exciting than the dreary reality of contemporary human life, where everyone looked, more or less, like their stomach ached. Myth brought sparkling Mediterranean seas and gleaming shrines to dolphins, where white columns floated in a cerulean sky. Myth brought dolphin stories told with flair and verve.

I found myself in Hawaii. A foundation had donated us a brand-new, wallowing ferro-cement ketch with the paint peeling off of its sides (some flaw in the construction), which had been built in Korea in a mistaken effort to enter the yacht market. The boat, which we named the *Tutunui* (great grandmother), was an exotic teakwood condominium inside, and a derelict outside, that we anchored in the warm protected waters at Kealakekua Bay in order to study the dolphins. My associates kept copious notes with compass points and times of "sightings" meticulously recorded. I learned to put on a mask without it leaking too much. We tried mightily to get the dolphins' attention: throwing frozen squid at them, dragging mylar kites through the water, tootling on plastic recorders, and blowing and burbling in our snorkels. They were not interested, swam by in lazy dim lines, wavering just on the edge of our vision, having something else to do. A friend tried constructing an alphabet out of naked human bodies—mostly airline stewardesses he picked up, who would do underwater acrobatics and spell out words like "hello" and "aloha." It was an inventive effort, and we all got better in the water, but the dolphins did not flock around our literacy program.

On one of my many journeys out of Lahaina, I rode a fine wooden ketch between islands and noticed a bottlenose dolphin riding the bow wave just below where I was hanging out over the bow. Terribly excited, I started talking in my mind to the dolphin, repeating endearments, being as loving and as seductive as I could manage, urging the dolphin to stay, and telling him or her just how lovely he/she was. For a long time we traveled together, above and below the bow, in some kind of kinship, and then I thought that maybe I should stop talking to myself like that, and the moment the sentence passed through my mind the dolphin sheered off and disappeared from my sight.

Myth moved close to reality. I camped on a cliff overlooking a small remote harbor and for over a year inhabited the place of my dreams. My home was a cement platform that had been used to evaporate sea water; it was sparsely furnished with a tent, an outdoor bed, three rocks for a fireplace, and a small cooler. My daily activities consisted of waking up with the sunrise, making coffee, adjusting the brightly colored sarongs that created my architecture and my shade, and watching for the dolphins or humpback whales that enlivened the waters at my feet. I overlooked a curved watery shelter in the lava cliffs, a distant volcano, which seemed to float on the sea like a great whale, and an expanse of ever-changing light and sky and water, replete with glittering stars, impossibly brilliant moonrises, and white frothy waves. I reveled in my surroundings, nestled deep into the arms of the ancient sea and sky, and felt close to the dolphins. At least we were in the same place at the same time.

The dolphins usually came into the bay beneath me in the early morning, swimming in big lazy circles, splashing and spinning and then disappearing beneath the surface, to rise and breathe and splash and swim. It was lyrical and calm to sit there, passive above the sea, and be near the dolphins. I rarely went searching for them, for I wished neither to disturb them nor to probe into their lives, but I would see pink-bellied babies, not too many at one time, and would see the family fishing: surrounding a school of akule or opelu by making a big net of bubbly splashing water to hold the fish in place. When a tour boat came into the harbor, the dolphins would often divert it by sending the bow-riders out. I would watch the boat come toward the breakwall and see the people standing on the bow and gesturing as the boat turned out toward sea again and carved circles in the water. Then, from behind the boat, where they had been resting all morning, the rest of the dolphins would resurface and swim and breathe. It was quite a fancy maneuver.

I tacked a small, hand-lettered sign to a keawe tree next to my tent that said "Dolphin Research Station," kept a large notebook with colored pencil drawings that noted weather, moon- and sunrises, and general information about where I was and what I saw, and learned, from the source I had sought, that everything in nature moves absolutely to the same rhythms. No longer was I an observer of the dolphins but just another fellow in the grand staging of life, and I was totally enchanted.

A few times, I went with a photographer to film the dolphins, and it was high comedy. We were loaded with gear and all kinds of stuff to keep us going, and the dolphins had, literally, nothing. We took the high-powered Boston Whaler along the coast, would spy the dolphins, drop the anchor, put on our fins and masks and snorkels, get our cameras, etc. etc., and jump into the water to watch the dolphins swim away. Then we would swim back to the boat, climb back inside, take off our fins and masks and snorkels, put on our sunglasses and hats and sunscreen, and roar off again and find the dolphins, and go through the whole routine again and watch them swim away.

Nature is illusive, she will not sit still, yet at the same time is as still as our perceptions might allow. Her shifts are subtle and hardly noticeable for the most part: a rock fallen here or there, a slightly different configuration in the shoreline or on the sloping sands. Time passes and passes again, and loiters at the horizon with the changing of the light from warm to cool and then to warm again. We are warts on nature, are not as patient as a tree frog, as withstanding as a mountain koa. We could not wait if our very lives depended on it. And they do. We humans are frippery—always in search of something—whereas animals seem settled, are willing to take naps, splash around in the sea, stretch in the sunlight, or rest after a meal. Animals are seductive; they lounge around, do not have jobs, do not have to lie. And many are very sexy. Dolphins especially. And dolphins became, for many, the promised grail, the way to innocence and love.

Scientists and others (me, too) wrote books and papers and speculated on the consciousness of dolphins, whales, and other animals; on how they might think, or if they thought. Was it truly possible that they were aware— conscious, like us, of where they were and where they were going? Could they plan? Could they remember? How did they meet? Could they possibly recognize each other or understand each other saying something in other than the king's English or French or Dutch? It certainly was possible. Remember that early ethnologists described the languages of others as gibberish, a language of apes or brutes. Slowly we have learned that those different from us do speak coherently, yet still, when confronted with another's language, we usually think of it as meaningless, a collection of sounds that has no authority.

A vast literature has grown, like mushrooms in the night, expelling everything anyone can think about dolphins and whales. Especially if it can be tested; if you can get tables of statistics or numbers from it. How do they find their way in the dark sea? How do they communicate? How big are their brains? How many cells in their cerebral cortexes? We sniff like hounds on the track, use sophisticated devices, all our tools, to probe into the being of another. It is no use. Nothing can ever come of it, because, you see, we cannot know like that.

We cannot know like that. We know, imperfectly, only with intuition, with the translucent mind of empathy and love. There is no other way we can know; the rest is mind-games, stuff to get grants for, to make careers and money out of, to exploit and use.

There are alarmingly attractive women and girls who have found no resonant echo in human men (who are too busy pursuing power or status to respond) and who search in the deep blue, sending out signals of love and wonder, in hope that someone will answer. Usually they hear no answer, so turn blindly toward another kind of illumination and seek out the dolphins. Dreaming of dolphins, they wear them imprinted on their breasts, dangling from their necks, caressing their wrists and fingers. Blinded by their desire, some prefer their dolphins in captivity, in tanks and research labs, where they can express their love and interest without the hindrance of the dolphins leaving them behind.

The quest is material nonsense because it cannot work and return the answer to that immense yearning, but it is a spiritual reality. Searching for love, and not finding it among our fellows, we look to find it in animals. Curiously, it arrives more often than not. An answer of sorts. It comes as a wagging tail, a tossed head, the sliding by of an eye in contact in a flurry of foam. Yes, we notice each other; indeed, we are both here.

Life in the sea is gestural; books get soggy in the surf and downright dissolved in the deeper water. There is not a whole lot you can do immersed in the ocean. If you are lucky, when you swim out to the dolphins they might wait, and then they might not. Glimpses are usually the most you get; a swish by and then nothing but blue-green water as far as your eye can see.

From a boat you can chase them around, a shoddy sport, and take notes or pictures. Just like anywhere else. You can learn the things that you define as knowledge, but you cannot learn what you do not already know, for there is no frame to the question, and even no question. I am told that now there is special software for dolphin studies, and if what you see or hear is not enterable on the spreadsheet, then it does not exist. Yes, you can do something from a boat, something different in a lab, but in the water we are less than equal, the dolphins and us. They can certainly swim away, and often do, leaving us lost and anxious in the channels between the islands, in the everlasting sea.

Forget the nonsense of how they think, and whether they are "intelligent" and have history or culture (our favorite subjects) or whether the males rape the females (another favorite subject). Forget about their kinship and their feeding habits, and whether or not they wean their young at six months or two years. Forget about "loving" them and making contact and learning their language. Forget about how far their voices travel, and whether or not it might be considered a song, a sound, or a language (as if there were an intrinsic difference). Forget everything we know—about us—and about all of the rest of us . . .

and walk from the dawn-washed shoreline into the welcoming sea

and swim . . .

KATHLEEN DUDZINSKI

Letting Dolphins Speak: Are We Listening?

Wow! They are so close. They encompass me within their group: one below, one on each side, others nearby. We swim in tight formation. They press their bodies to mine! Simple contact as we swim against a slight current. Thrilling, exhilarating. Words are not enough to describe the feelings—but, wait. It also feels odd, off from what I remember.

Five years ago, members of this group of Atlantic spotted dolphins rarely touched swimmers. Sure, individual dolphins approached within a couple of feet of us, but their familiarity, their acceptance of swimmers now seems drastically increased from just a few years previous. What does this mean to those of us attempting to study dolphin-to-dolphin relations, behaviors, communication, and overall activity? The situation of research on wild dolphins involved in swim-with-the-dolphin programs is a catch-22. We attempt to be as noninvasive as possible. We attempt not to disrupt the dolphins' natural activity. After all, our goal is to better understand the normal, uninterrupted behavior of wild dolphins. But, for about 85 percent of the studies on wild dolphins, we cannot shroud our presence. We are on boats or underwater. We cannot hide.

This is the paradox of studying dolphins associated with swim-with-the-dolphin programs. We, as scientists, prefer to be ignored by dolphins, to remain as a "fly on the wall," so to speak. Participants in the swim programs, however, are there for precisely the opposite reason—to interact, to play, to join in the dolphin fun. How do we assess the difference between non-interactive, noninvasive observations and swim-program participants with a desire to join dolphin society even if only fleetingly? It is nearly impossible, and the only behavior we can hope to directly control, or modify, is our own and that of the human swimmers.

ABOUT DOLPHINS

Let's back up a bit. What are dolphins? What do we know about these marine mammals? Dolphins are herding animals that live within complex social groups. Living in groups provides a number of benefits to individual members, including easier detection, location, and capture of food, avoidance of predators (that is, not becoming food), resource defense, reproduction and mating, and care of offspring. Communication between animals is a prerequisite for coordinated social organization. Communication can be accomplished through visual, auditory, and tactile sensory modes: signals can be exchanged among individual animals along each of these sensory channels. Signaling behavior is often dependent upon context (for example, type of environment, presence of different friends). Dolphins employ complex social signals and display a variety of behaviors: these characteristics facilitate many types of coordinated interaction. Signals help group members communicate more effectively during different activities, including courtship, warning (or alarm), and the establishment and maintenance of social relationships and hierarchies.[1]

Because their lifestyles are completely aquatic, dolphins rely primarily on sound or information (and signal) exchange, particularly in murky waters. Dolphin sounds are broadly divided into three categories: whistles, clicks, and other burst-pulse sounds. Burst-pulse and click sounds (short in duration with energy across a large frequency range) are likely used by dolphins to orient and navigate in their surroundings and to search for prey. Although we have little evidence or hard data to support the possibility, click trains may be used by dolphins to communicate about food or their environment. Or, maybe about some topic we, as land-dwelling creatures, have yet to understand. Dolphin whistles sound quite similar to human whistles, but dolphin whistles are slightly higher in frequency (pitch) than human whistles. Most researchers believe that dolphins use whistles primarily to communicate about their social setting, activities, and relationships. It would be difficult at this stage, if not impossible, to determine whether dolphins use syntax in the production and use of their sounds.

Still, communication is achieved by more than just sounds, and this presents a much easier subject for study and consideration in wild dolphins. In clear water, visual signals, such as dolphin coloration patterns and posturing, may play a more prominent role in communication among individuals than in murky waters. How dolphins use vocal, visual, and touch signals varies according to individual, sex, age, associates (friends or foe), and behavioral context, as well as by environmental factors. My research objectives focus on attempting to understand possible correlations between the sounds and behaviors of dolphins, including any variation among individuals.

DOLPHIN-DOLPHIN INTERACTIONS

Studying dolphin-to-dolphin behaviors and relationships provides us with a better understanding of how and why dolphins interact socially. Details of their reactions to and initiations of specific behaviors with particular associates sheds light on their use of signals, as well as the potential meanings and messages in their cues. We need a strong understanding of dolphin-to-dolphin behavior (that is, what is acceptable in dolphin society as "good manners") to begin to understand how human presence may affect their behavior, relationships, and distribution.

I study how dolphins interact and how they communicate, and I have come to believe that dolphins do not use a language like ours: a language filled with words and grammar and syntax. Rather, dolphins use a complex web of signals to communicate with each other. Dolphins share information not only through sound but also, and especially, by touch. While traveling, resting, socializing, or playing, dolphins are often seen in physical contact with other dolphins. Touch can occur with almost any body part, but some areas seem to have special significance. For example, when one dolphin places its pectoral fin (flipper) against another dolphin's side between the dorsal fin (curved fin on its back) and the tail, this is a request. The actual favor being asked depends on the context, the ongoing activity, and the individuals involved. Nonetheless, the touching dolphin initiates, or sends, the request with its flipper. A limited list of other dolphin signals, with possible meanings, is given in Table 1.

For communication to occur, three things are needed: a signal, a sender, and a receiver. A signal can be anything perceived—a head nod, a sound, or a touch of some kind. A signal is the *vehicle* for communication to occur

Table 1
Dolphin Behaviors and Possible Meanings in Human Terms

We have yet to find the Rosetta Stone to help us communicate directly with dolphins verbally. There are some recurring themes in dolphin behavior, however, that lead us to certain assumptions regarding their interactions. This table includes some of the more common signals and their potential meanings. Many of our inferences about meaning depend on the context in which a signal is sent.

SIGNAL	POTENTIAL MEANING
Bodily Contact	
Body rubbing	Affection or affiliation, strengthening social bonds, reaffirming relationships, quieting an excited peer (usually a youngster).
Flipper-to-flipper contact	A greeting between two dolphins.
Flipper-to-side contact	A request for a favor or help sometime in the near future.
Hits, rams, slams, bites	Usually irritation or aggression from older dolphins, but playful when accompanied by soft angles of approach.
Melon-to-genital contact	When a mother and calf are swimming in echelon, the calf will often touch mom's genital area with its melon, maybe indicating it wants to nurse (i.e., is hungry). If the mother initiates contact, then maybe it is telling the calf to nurse.
Vocalizations	
Whistles	Often dolphins produce a stereotypic whistle that is usually called a "signature." We now know this is used for maintaining contact among dolphins.
Chirps	Short sounds that resemble bird chirps, these may signal a dolphin's emotions, sort of an "okay" message.
Click trains	Short pulsed sounds of high, wide-band frequency that are used to investigate objects or search for fish. Often these sound like creaky old doors opened slowly.
Squawks	Pulsed sounds like "squawks" and very high in repetition rate. They are used mostly in fights or during play and may signal anger.
Bubbles	
Bubble stream or trail	A stream or trail of little bubbles that escape from a dolphin's blowhole, often seen from young dolphins during excited and playful swimming.
Bubble clouds	Large pockets of air from the dolphin blowhole ("clouds") that may be used to express anger or warning. They may also be used as a "shield" from another dolphin's harsh vocalizations.

and contains two parts: the message, and the meaning. The *message* is the signal the sender transmits, but the *meaning* of a signal is what the receiver gets or understands. These two parts are often not exactly the same because of the differences in context, or rather history, for both the sender and the receiver. Confusing, I know, but let me give you an example. Walking down a busy street, you see a mother and child on the opposite sidewalk. Suddenly they grab hands. Why? Many explanations pop to mind—danger crossing the street, excitement at a window display. Now, if I tell you the mother and child are in a large park with lots of birds and trees, but no cars, the possible meaning of their "hand grab" could be very different. Context includes not only the environment but also the ongoing activity and relationship between sender and receiver. Similarly, a signal's message and meaning depend heavily on who is the sender and who is the receiver. In our example above, if you know the mother grabs the child's hand on the busy street sidewalk or the park, then your range of possible signal meanings is different from if the child grabbed the mother's hand.

This is how we study communication, how we learn what messages dolphins are sending and receiving. We watch not only *how* they interact but also what *roles* each individual assumes and what the *context* (surrounding activity or setting) includes. (This is why studying communication of other long-lived social animals takes so many years.)

MIKURA ISLAND DOLPHINS

My studies include not only the Atlantic spotted dolphins in the northern Bahamas but also bottlenose dolphins found around Mikura Island, Japan. Work on these bottlenose dolphins offers a unique opportunity to understand dolphin behavior in a setting far different from the Bahamas. Mikura Island is a dormant volcanic island with a boulder coastline. The ocean near shore ranges in depth from four to twenty meters. Dolphins can be seen in the surf zone (within two meters of the coast) near Mikura Island as well as in the deeper waters offshore. In the Bahamas, spotted dolphins are regularly observed between forty and sixty kilometers from shore in shallow water (less than eight meters deep) with a white sandy bottom. Both locations, however, offer clear water and a tolerance of human swimmers by dolphins to give us predictable opportunities for observing dolphin social behavior and communication from underwater.

Although spotted dolphins in the Bahamas have swum with people since the late 1960s, it was not until the spring of 1994 that dolphin-watching and swimming programs began with a vengeance around Mikura Island. The coastline of Mikura is far more accessible to people than the sandbar north of Grand Bahama Island. The numbers of visitors to swim with Mikura bottlenose dolphins has steadily increased; for example, from May to September 1998, more than 10,000 tourists took a trip to swim with or watch dolphins around Mikura. That is at least twenty-five times the number of people who accompanied me on research trips to the Bahamas for four years, six months a year! Competition for tourists as well as for the dolphins is intense and occurs both within and between businesses and boats. Without regulation, these hordes of people have the potential to actually do more damage than good to dolphins—in essence, "loving them to death." Luckily, the Fishermen's Cooperative Association, dolphin guides, and boat captains collaborated to establish guidelines and rules for these interactions. Dolphin-swim programs provide a strong economic base for a small island, yet how can we be sure we are not ultimately changing the normal behavior of these dolphins? We may never find the answer to that question, though we occasionally get glimmers of insight.

I spent two years living in Japan studying these bottlenose dolphins, catching rides on fishing boats for the dolphin trips whenever there was space. Although I do not study human-dolphin interactions, I could not help but witness the aggressive tendencies of the swimmers toward the dolphins. Mostly people thought they could catch a dolphin and, once caught, the dolphin would just love to be touched all over. Imagine having a stranger walk up to you with their hands extended in an effort to touch you! Amazingly, many of the swimmers did not understand the effect of what they were doing. (The same is true for participants traveling to swim with spotted dolphins in the Bahamas.) The desire to touch and be close to a wild animal that seems to accept us is a powerful, at times mesmerizing, attraction—one that may inadvertently lead to a serious miscommunication between species. If the dolphins become accustomed to a specific type of behavior from us "two-leggeds," then they may solicit it from those of us acting differently, albeit respectfully. My return to Mikura Island after a six-month hiatus suggested that these dolphins were becoming quite bold with swimmers.[2] Had these dolphins come to expect more aggressive tendencies from humans?

To many researchers, bottlenose dolphins have a reputation as bullies, with evidence of infanticide from Scotland as well as their being implicated in the deaths of several porpoises.[3] Similarly, several cases have been reported of aggressive behavior by lone, sociable bottlenose dolphins directed at human swimmers.[4] (I suggest that these aggressive actions are "trained" responses to the repeated actions of numerous swimmers.) Personal observations during the summer of 2000 indicated to me that at least some of the Mikura dolphins may be traveling the same path. The dolphins had become more aggressive and direct in their actions toward swimmers. Head-on swims and threats to people were more prevalent than previously. Interestingly, one observation suggested that dolphins may also have learned how to use people as a type of tool. Let me describe what we witnessed.

We were observing from the surface a small group of dolphins in what appeared to be low-level social activity: some splashes, leaps, and frolicking. Our boat was within eight meters of the coastline. We entered the water to observe the dolphins more closely and heard loud, intense whistling and sharp pop sounds about fifteen seconds before seeing three dolphins. (I should interject that there were five people in the water: two of us recording behavior for our research, and three participant swimmers. We were spaced out in the water and all watching this activity.) As it turns out, there were two adult male dolphins attacking a subadult female. She was whistling repeatedly, with a stream of bubbles from her blowhole, and had fresh bleeding scars from the males' teeth. In human terms, it seemed blatant to us that she was whistling for help. Or maybe trying to "talk sense" into the males? The males were persistent and likely attempting to corral the female for mating purposes—bottlenose dolphins have been observed herding females in Sarasota Bay, Florida, and in Shark Bay, Monkey Mia, Australia.[5] (In fact, mating was witnessed about ten minutes after our observation by another researcher.) I was stunned when the female swam away from the males and in among all of us humans; we had inadvertently formed a sort of circle as we watched the activity. Was her persistent whistling directed (sent) to us? If so, we missed the message entirely. Why were other dolphins not coming to the aid of this female? If only we could decipher her signals more specifically. But, even if we had, how would we have offered assistance? Maybe the female realized this and took what she needed by using us as a shield, a tool? After all, the males remained about eight meters from us, all the while posturing aggressively and producing intense, loud pop sounds.

After sixty seconds the female sped away, with the males in hot pursuit. Had she gotten what she was whistling for? Were we the intended recipients? Or, along another line of reasoning, could she have been signaling to the males to "back off" for a bit of time, to give her a reprieve? I feel the latter is unlikely, but we are hardly in a position to judge meaning in dolphin signals with conviction. One day maybe . . . but for now we have only mildly educated guesses.

This observation is important for two reasons (not including the amazing behavior and interaction we observed between the dolphins). First, the female dolphin purposefully positioned herself among the humans. She surrounded herself with us. It seemed to all humans involved that we provided a brief reprieve for her from the attacking males. But is this truly what she intended to do? And how would her behavior have been different had we not been present? We may never have the answer to this last question. Second, the other researcher and I immediately recognized the gravity of our situation and the danger we were in had the males decided to remove the obstacles—us—in their path to the female. The three participants, however, had not a clue of the danger they were in. Indeed, they described the situation as cute and fun; they did not even realize the female was being attacked! This is an example of extreme miscommunication—explained by the fact that these three swimmers did not have background information on dolphin behavior. Their contextual information or setting was limited by lack of knowledge about their given circumstances. The three humans' reaction to the behavior of these three dolphins simply drove the point further home for me. We must provide useful, strong, and wide-reaching education programs to teach people how best to interact with wildlife, especially dolphins. We are not going to stem the tide or the desire people have to interact personally with dolphins. But we can make every effort to ensure that people act responsibly and are informed of proper etiquette when near wild animals.

HOW DOLPHIN-TO-DOLPHIN COMMUNICATION RELATES TO DOLPHIN-HUMAN INTERACTIONS

Over at least the past fifteen years, human interest in dolphins and whales has increased at an exponential rate. This trend shows no sign of declining. With our heightened interest has come a desire to observe and interact with dolphins and whales in their natural environment. A positive effect of this interest is a renewed desire to protect the environment. People will seek ways to change their relationship with nature. Eco-tours have become popular vacation venues for many people, and arriving at a compromise between passenger expectations and actual experiences can be challenging. Participants need to be taught how to behave responsibly with their environment so that they better understand the rigors and requirements of scientific (and especially behavioral) studies under natural conditions. Similarly, participants need to understand how research and management issues complement one another. And it is my strong belief that these issues must complement each other for the protection of our natural resources.

The study of dolphin-to-dolphin communication will shed light on what behavior is acceptable in dolphin society. Dolphins are social mammals and show teaching and observational learning. They must, like other social mammals, learn which signals are appropriate given specific situations. If we, as observers, can get a clearer understanding of what signals dolphins use, and when, then we can better direct our behavior when interacting with and observing dolphins.

WHAT WE CAN DO

Currently, I split my summer field seasons between the two study sites and dolphin groups, continuing my research into their communication. The dolphins in each group have become a surrogate family to me—at least through the photographic catalogs I keep to identify individuals. Over ten years in the Bahamas and six in Japan, I have watched as the young mature and learn about their society. I learn by bits with them. Cultures are not static but change and adapt with generations. This is true for humans and for other social animals. Dolphins are no exception. Thus, I cannot help but wonder whether the generation of spotted dolphins that grew up with swimmers has so completely incorporated us into their culture that their offspring are taking it one step further. You see, the young spotted dolphins that encompassed me and exchanged rubs with me upon my return to the Bahamas were the offspring of the dolphins I have watched mature and grow. We may not be able to direct dolphin behavior in the wild, but we can direct human behavior. We can attempt to make our actions as nonintrusive as possible. My studies of dolphin communication have led me to several conclusions about how we can best interact with them and have prompted me to write what we refer to as "codes of conduct" for people who interact with dolphins.

It is important to realize that the ocean is not our home but our playground. It is, however, home to dolphins, whales, and numerous other marine organisms. We are guests and should act accordingly and with caution. Though equipped with a huge smile and a seemingly forever-harmonious disposition, dolphins and whales are wild animals and should be treated with respect. If our future with dolphins is to include real interspecies communication, respecting dolphins enough to "listen" to them would be the first step.

NOTES

1. B. C. R. Bertram, "Living in Groups: Predators and Prey," in *Behavioural Ecology: An Evolutionary Approach*, ed. J. R. Krebs and N. B. Davies (Oxford: Blackwell, 1978).

2. In August 2000, the volcano on Miyake Island (just north of Mikura) erupted, causing an island-wide evacuation. This drastically reduced the numbers of tourists coming to swim with this group of bottlenose dolphins.

3. I. A. P. Patterson, R. J. Reid, B. Wilson, K. Grellier, H. M. Ross, and P. M. Thompson, "Evidence for Infanticide in Bottlenose Dolphins: An Explanation for Violent Interactions with Harbour Porpoises?" *Proceedings of the Royal Society, London, B* 265 (1998): 1167–70.

4. K. M. Dudzinski, T. G. Frohoff, and N. L. Crane, "Occurrence and Behavior of a Lone, Sociable Female Dolphin *(Tursiops truncatus)* Off the Coast of Belize," *Aquatic Mammals* 21, no. 2 (1995): 149–53; M. C. O. Santos, "Lone Sociable Bottlenose Dolphin in Brazil: Human Fatality and Management," *Marine Mammal Science* 13, no. 2 (1997): 355–56.

5. J. Mann, R. C. Connor, P. L. Tyack, and H. Whitehead, *Cetacean Societies: Field Studies of Dolphins and Whales* (Chicago: University of Chicago Press, 2000).

Dolphin Healing from Afar

M y dolphin trail began nearly thirty years ago, when on vacation off the Isle of Man, in the Irish Sea. I was on a scuba-diving holiday with a close friend. We were both heads of different medical research laboratories, studying the brain and the central nervous system. Little did I know that this holiday would provide an unforgettable inspiration for new research and something that would completely change my life.

In the harbor of Port St. Mary, we met a dolphin who had become friendly with the local divers. He was a large, mature male bottlenose dolphin (*Tursiops truncatus*) and was referred to as "Donald." When I had previously founded the Oxford Underwater Research Group in 1963, one of our aims was to find ways to reduce some of the hazards of scuba diving. During their evolution from land mammals, dolphins had overcome the problems we faced and become superbly adapted to life in the sea. They did not suffer from the bends or nitrogen narcosis. Furthermore they had brains that were as large, more highly evolved, and some thirty million years older than those of humans. So when I first heard about Donald, I immediately started to think of the dolphin as a potential experimental animal. However, that attitude was dispelled the minute I got into the water with him. Instead, *I* felt like the experimental animal. All thoughts of scientific research flew out of my head as we started to play together in the water. Life was one long game for this dolphin.

The story of how my relationship with Donald unfolded was the subject of several books[1] and a major documentary.[2] The television film recorded the remarkable relationship I was able to establish with Donald, who was about three and a half meters long, weighed approximately 250 kilograms, and was totally free to swim away at any time. What was extraordinary about our friendship was that it was *his choice*. There was no way I could swim after Donald if he decided to go somewhere else—which he often did. This gave rise to an interesting change of attitude on my part with regard to the studies I tried to conduct with him. I had to accept that it was the dolphin who was in control—not me.

One outcome of meeting Donald was that I gave up orthodox medical research and devoted myself to finding out what it is about dolphins that give them such a special place in our hearts and minds. Toward this goal I founded the nonprofit International Dolphin Watch (IDW) in 1978. The main aim of IDW was to observe and protect dolphins and to gain greater insight into their behavior in the wild. My books[3] and films,[4] together with the rapid growth of sports diving in the 1970s and 1980s, have resulted in many more encounters with dolphins in their natural environment.[5] Some dolphins are taking the initiative and following Donald's example. They deliberately seek out the company of divers. This enables studies of free wild dolphins to be conducted in ways that would have been impossible before. We now refer to such dolphins as Ambassador Dolphins. These Ambassador Dolphins are directly responsible for strengthening the human-dolphin bond that has existed for over 2,000 years but was confined to surface swimmers in earlier times.

My work—which is really better described as play—with friendly, solitary dolphins led to an amazing discovery. When the film I helped to make about a dolphin we called Percy off the coast of Cornwall in England was shown on television,[6] the BBC was inundated with mail. Many of the letters were from viewers who suffered from depression. They reported that the film had lifted their spirits. Soon afterward, IDW initiated a research project entitled Operation Sunflower. The seeds for this were sown when my companion commented that a man who had suffered from chronic depression for twelve years had "blossomed like a sunflower" following an encounter with a wild dolphin. The aim of the investigation was to see if dolphins could help those afflicted with clinical depression.

The research was conducted and filmed with a solitary dolphin off the coast of Ireland.[7] The data we gathered did not fulfill the strict criteria for a proper clinical trial. Nonetheless, it provided convincing evidence that dolphins could help people diagnosed as depressives.

But the success of this study posed a major problem. An estimated one person in ten in the Western world is expected to need some form of psychiatric help during his or her lifetime. Thus, in Britain alone, there were potentially five million people who might want to swim with a dolphin. Taking even a tiny fraction of these people out into the sea to swim with dolphins was clearly impossible. A search was therefore started to find a method of reproducing the healing essence of dolphins. The conventional way of doing this was to produce a dolphin "pill." But because no direct chemical interactions were taking place, what possible form could such a pill take?

The Australian Aborigines had the answer.

I discovered that, with the aid of music, they could take a listener into a mental state akin to that of an encounter with a dolphin. I was also fortunate enough to hear a recording, made in Australia, called *Dolphin Dreamtime*. When I first listened to the tape, in the dark while floating in a pool, I had an extraordinary, out-of-body experience. I had the sensation of water flowing across my body as I swooped into the depths. I was a dolphin. I knew what it was like to be a dolphin. The heavy feeling that had prompted me to go for a swim evaporated in a flash.

So I had my pill—an audio pill—that would take the listener on a sound journey into the world of the dolphins.

Through IDW, the *Dolphin Dreamtime* cassette was made available on a random basis. Each cassette was accompanied by a questionnaire that was returned to me in confidence. Analysis of the first responses by the Applied Psychology Unit of the Medical Research Council in Cambridge in 1990 was sufficiently encouraging for the trial to be extended. A twelve-page statistical analysis conducted by Richard Pearl in the Department of Psychology at Swansea University of the 173 responses received up to 1994 indicated that over 70 percent of those listening to *Dolphin Dreamtime* benefited from the experience.

Dolphin Dreamtime (now also available as a CD from IDW, www.idw.org) has become established as a useful tool in psychiatric wards. In addition to helping those diagnosed with clinical depression, *Dolphin Dreamtime* is finding ever-widening applications. These range from post-operative trauma to tension release in prisons. A physician uses it to help him deal with his own disability: myalgic encephalomyelitis. Another doctor listens to *Dolphin Dreamtime* to counteract her insomnia—especially when she is on call during the night.

Listening to the tape or CD is a very benign form of treatment. So far no unpleasant side effects have been reported. Children respond especially well to *Dolphin Dreamtime*, and many now use it for the release of tension before examinations.

Probably because they are so spontaneous and playful, dolphins have an extra special relationship with youngsters. This became abundantly clear to me in 1974, when I had the truly magical experience of seeing Donald scoop my thirteen-year-old son Ashley onto his head and give the boy a ride around the harbor of Port St. Mary on the Isle of Man. Ashley was jubilant. When he told his friends, they were amazed. It changed his life. It made him even more aware of the importance of caring for wildlife—especially dolphins.

That experience led me to the idea of creating a make-believe dolphin that would appeal directly to children and the childlike spirit that most adults harbor but do not always admit to. The advantage of this strategy is that I can put the dolphin in situations, based on real-life experiences, that show the joy they could bring to humans. At the same time, it can reveal their vulnerabilities and how important it is for humans to care for dolphins and treat them respectfully.

I chose the name Dilo for my fictional dolphin. This made-up name I hoped would be acceptable in all parts of the world, especially in countries where there was scant knowledge of dolphins and little regard for their well-being. Through my Dilo stories, I also want to raise awareness of the need to keep the oceans of the world in a healthy state in order that dolphins and all other sealife can survive and flourish.

My books about Dilo have been translated into many languages, including Chinese and Japanese. Through Dilo I encourage young people around the world to become Web Warriors. Using the World Wide Web and e-mail, even severely disabled children can lobby politicians and demand changes in international laws that will protect dolphins.

In the first book,[8] Dilo's mother brings up her baby alone, and they become best friends. The young dolphin's curiosity gets him into many amusing situations. After his mother dies, caught in a fishing net, Dilo knows her spirit is always around him. At night he sees her outlined in the stars.

Sally Galotti, an exceptionally gifted artist who has worked for the Walt Disney Organization in Italy, the United States, and England, has illustrated two of my Dilo books.[9] One of our aims is to make a full-length movie with Dilo as the central character. However, the BBC is now looking into the possibility of creating an animated cartoon series about Dilo for television.

When Sally painted pictures of Dilo on the walls of a hospice in Romania, where the children were dying from AIDS-related illnesses, both the staff and the patients showed astounding positive behavioral responses to her joyful images. The effect was so profound that Sally decided to dedicate the next years of her life to applying her consummate artistic skills to helping sick children. She is now decorating the walls of children's hospitals throughout Italy with pictures of Dilo on the basis that the sense of fun and release from fear and stress they engender make young patients more responsive to treatments of all kinds.

In ancient Greece, it was believed that drowned sailors were transformed into dolphins and that dolphins carried the souls of the dead into the next world. Having seen the benefits that Dilo can bring to dying children, Sally and I are now working on ways in which Dilo can help terminally ill children and their loved ones come to terms with death. The most powerful image we have produced in this respect shows Dilo at night looking up into the sky and seeing his mother outlined in the stars.

The use of stories and images to depict the unique connection between humans and dolphins is not new, of course. Many of them have been passed down through the generations. One such story is told by the Chumash Indians, who lived on the coast and offshore islands of south-central California. They were told not to look down when they crossed the Rainbow Bridge to the mainland. Those who could not resist the temptation fell off the bridge into the sea, where they were transformed into dolphins.

The legend of Arion being saved by a dolphin and ferried ashore is depicted in a sculpture made about 2,500 years ago. The story was reported by the Greek historian Herodotus in 450 B.C. and has been retold many times since. Recently it provided the basis for an opera, *Arion and the Dolphin*, performed by the English National Opera.[10]

Sally is also working with me on the Dilo Dolphin Dome Project in association with Trevor Goldsmith, fisherman, sculptor, and genius designer of high-speed boats. Between us we are using art, science, and the latest technology to create interactive dolphin-like experiences from film of wild dolphins and fantasy cartoon images of dolphins. The aim of the project is to bring joy and healing into human lives—especially to children with severe disabilities from deprived backgrounds. It is our intention to build a Dilo Dolphin Dome that can be disassembled and taken to different sites around the world where there are children with special needs, especially in impoverished countries like Romania.

I now propose to build on these achievements and extend the research of Operation Sunflower with an investigation into the role dolphins can play in improving literacy. I do so on the basis that learning a language involves developing and reinforcing mental processes that can be stimulated and improved by real and imaginary dolphins. The idea for conducting this research arose from the numerous unsolicited letters I have received, mostly from highly literate children, telling me how much they have enjoyed the Dilo Stories. Mixed with them were reports that Dilo's antics were giving children with learning difficulties encouragement to read.

This observation was reinforced by the Dilo Publication Project 2000, which culminated in publication of *Dilo and the Treasure Hunters* with the help of a group of 9th grade students.[11] When the students took copies of the book to a local elementary school, they successfully stimulated 4th grade pupils of varied abilities willingly to read out loud. This caused me to wonder if this was achieved as a result of the enthusiasm of the high school students alone, or did the fact that Dilo was a dolphin play a role? I now believe that it was a combination of both, acting synergistically. I realize that quantifying the contribution of the two separate components (the teacher and the teaching material) would be difficult. Nonetheless, I drafted a research protocol.

Hazel Smelt, a special educational needs coordinator, undertook a pilot study with 7th grade students with reading difficulties. It was an immediate success. So much that students who were not involved wanted to know why they were excluded. They asked for Dilo to be introduced into their English lessons.

Since then a full research program, the Dolphin Education Research Project, has been set in motion. The aim of this project is to provide sound scientific evidence to support the introduction of dolphin stories and dolphin-related activities into the teaching of English. The ultimate objective is to help children of all abilities to attain their full potential—especially those who find learning English difficult. In addition, the project aims to create awareness of dolphins and their environment and the need to protect them. This in turn has led to a holistic approach to education in which I have taken my lead from the dolphins. Namely, that learning should involve parents and be fun for the students and their tutors.

My latest book[12] has taken me back to my medical/scientific roots. In it I have analyzed the evidence for the power of dolphin imagery to stimulate the human mind. I have paid particular attention to the so-called Indigo children, who want to change the world and can be very disruptive. I also look at alternative ways of managing the five million children in the United States who are medicated daily with the drug Ritalin (the long-term side-effects of which are unknown) to control their attention-deficiency disorders.

When I met Donald in 1974 and gave up orthodox medical research, I had no idea where my search for the magic of dolphins would lead me. It has taken me around the world many times. I have journeyed on a square-rigged sailing ship down the Barrier Reef in Australia where I have been with dolphins and humpback whales. My quest has led me into the mysteries of traditional Chinese medicine and life in a Buddhist temple in Kyoto, Japan. Where it will go from here I do not know. It seems my dolphin trail still has a long way to go. But with an awakening awareness of the power of dolphins to heal from afar, the limits seem boundless.

NOTES

1. Horace Dobbs, *Follow a Wild Dolphin* (London: Souvenir Press, 1977; rev. ed. 1990); *Save the Dolphins* (London: Souvenir Press, 1981); and *Follow the Wild Dolphins* (New York: St. Martins Press, 1982).

2. "Ride a Wild Dolphin" (YTV, UK); available from International Dolphin Watch, www.idw.org.

3. Horace Dobbs, *Classic Dives of the World* (Sparkford: Oxford Illustrated Press, 1987); *The Magic of Dolphins* (Guildford: Lutterworth Press, 1984); *Tale of Two Dolphins* (London: Jonathan Cape, 1987); *Dance to a Dolphin's Song* (London: Jonathan Cape, 1990); and *Journey into Dolphin Dreamtime* (London: Jonathan Cape, 1992).

4. "A Closer Encounter" (Channel 4, UK) and "Bewitched by a Dolphin" (HTV, UK).

5. Melanie Parker, *At-a-Glance Guide to Where in the World You Can Have a Dolphin Encounter* (North Ferriby: Watch Publishing, 2001).

6. "Eye of a Dolphin" (BBC TV, UK).

7. "The Dolphin's Touch" (TVS, UK).

8. Horace Dobbs, *Dilo and the Call of the Deep* (North Ferriby: Watch Publishing, 1994).

9. Horace Dobbs and Sally Galotti, *Dilo e il richiamo degli abissi* (Vicenza: Edizioni il punto d'incontro, 1997); Horace Dobbs and Sally Galotti, *Potere magico di Dilo* (Milan: Parole di Cotone, 1999).

10. Horace Dobbs, *Dolphin Healing* (London: Piatkus, 2000).

11. Horace Dobbs, *Dilo and the Treasure Hunters* (North Ferriby: Watch Publishing, 2000).

12. Horace Dobbs, *IDEAL (Integrated Dolphin Education and Learning): A New Gateway to Literacy* (awaiting publication).

With a Little Help
from My Friends

Sometimes I wonder what it is like to be a dolphin. How is living in liquid different from our own orientation in air? Is my perception of reality—based on what I see and hear and feel—different from the reality construct of a dolphin, who senses by echolocation, the practice of using sound waves to decipher the environment? If I were a dolphin, could I pass sound waves into my partner's heart and learn not only about his clogged arteries but also about his love for me? Could I send out vibrations to keep track of my teenage sons, miles from home on a Saturday night? Could I invoke the power of water to aid me in my work as a healer?

Like a dolphin, I have honed my senses in order to improve my interpretation of the world around me. I have been a chiropractor for over eleven years; in chiropractic school I was trained to distinguish separate spinal bones or vertebrae by touch. These many years later, I can feel through several layers of human tissue; sometimes I am able to feel *through my hands* the emotional release of trauma stored in the body. I have refined my sensory skills according to the principles of CranioSacral Therapy (CST) and Somato Emotional Release (SER). Developed by osteopathic physician John Upledger, these healing techniques focus on releasing connective tissue, especially around the brain and spinal cord, thus improving performance of the central nervous system. This teaching model uses human anatomy as its base, but often it seems that the human psyche plays an equally important role. Particularly striking are the effects that CST and SER have on the human mind and body when the therapy is combined with time spent in the water—and with dolphins.

In my healing work, I have come to believe a theory first proposed to me by Upledger, one based on the idea that dolphins possess healing abilities that can be shared with humans. I did not believe, however, that my skills were sufficiently developed to work with them. A surprising experience showed me that perhaps dolphin energies are like radio waves: available to anyone who can listen.

I was treating Matt, who had come to me for help with headaches and digestive problems. He sat on the futon-covered treatment table, methodically reciting his history. The highly structured training of an academic life had taught him to approach health problems with the same rigor as he would any research project. The man had done his homework. Medical consultation, complete with MRI and colonoscopy, ruled out any disease that could be cured by medication, radiation, or surgical excision. Matt had learned what not to eat, and although he was not always successful in changing his behavior, he was learning to recognize when his behavior triggered emotional stress in his body. He was also recognizing connections between his digestive problems and the frequency of migraines. Matt was seeing an acupuncturist and meditating regularly.

I gazed at this slender, bearded man. His blue eyes squinted back at me, chin slightly lifting. My comfortable office did not put him at ease. The old fir floor and high ceilings that I love so much did not invite him to breathe deeply and relax. He did not surrender the tension in his throat, neck, and shoulders to the warm embrace of desert wind, woven into the tapestry on the wall. Even the tumbling of water over rocks in the fountain soothed only me. Here was a man fluent in the language of holistic health yet apparently unable to embody it.

"The secret of CranioSacral Therapy," I said, "is to remember that I am not going to fix something that is wrong with you. My intention is to help you listen to what your body is trying to tell you." Demonstrating with my open hands opposite one another, as if cradling his body between them, I explained, "With gentle touch to your sacrum, spine, and bones of your skull, I will feel the quality of the connective tissues, especially around the brain and spinal cord, following any releasing your body initiates. It is very subtle and will probably feel soothing."

I did not explain how the nerves supplying his digestive tract exit the spinal cord from the mid- and lower back bones, or that any restriction in the spine, or in the connective tissues of the back bones, could compromise nerve function to the intestine, causing him inflammation and pain. Neither did I explain the complex relationship of cranial bone movement to circulation of both blood and cerebrospinal fluid throughout the head, neck, and spine, or how interference in this system could be a factor in chronic headaches. I did not want to give him information that might further alienate him from the sensations in his body.

I invited Matt to lie face-up on the table. I began with hands lightly on his outer ankles. I was using the simple CST technique of "arcing" to determine where to start. Arcing is the practice of feeling vibration through the tissues of the body. CST teaches us to recognize the body as water, more fluid than solid. We use CST to find scar tissue, old injury, disease process, or even stored emotional trauma. Any of these conditions can cause a disturbance in the otherwise harmonious balance of the body. These disturbances can actually be felt by a trained practitioner. Like a rock thrown into still water, any physical disturbance sends concentric ripples of vibration outward. The effect may be as subtle as a note out of tune or as dramatic as a huge fallen tree jamming a rushing river. By moving hands in the direction of smaller and smaller ripples, one is drawn toward the part of the body most needing attention.

But with Matt, I was surprised that I could not feel anything. Matt's body seemed completely closed to me. Not only was I stopped by the postural guarding of tight muscles, but it was as if he had put up a wall to block me from feeling any movement at all through his tissues. The next moment it felt as if I were being bombarded by a confusing sea of ripples, restrictions fighting for my attention, none reaching my hands clearly enough for me to follow.

I sighed deeply, not knowing how to proceed. I must have slipped into an altered state, because suddenly mental images of swimming dolphins startled me. I felt strong currents of flowing movement in Matt's body. I knew Upledger had successfully requested "dolphin assistance" during treatment sessions, but I had not actively visualized dolphins. These dolphins seemed to show up unannounced. Perhaps it was I who was the guest. It was as if they were already present with Matt's healing process, and by "listening" to his body with my hands, I stumbled upon them. My hands felt as though they too could swim, accompanied by these dolphins through the inland sea that was Matt's body. My own brain felt unfettered, much like the welcome relief of a deep meditation. I was free to observe myself as capable healer, insecure human, and curious interspecies traveler. I wondered if other therapists who reported treatment sessions with "dolphin assistance" had similar experiences. Could it be that what I was perceiving as "dolphin assistance" was more a shift in my own consciousness, allowing me to "go to" a place of increased awareness and perception, shared by dolphins? Perhaps this was related to increased sensory perception in my hands, akin to the vibratory sense of echolocation.

Matt's body came alive to me as my entire body became receptor for movement beneath my hands. My hands listened through layers of Matt's body as if sifting through sand, looking for a lost coin, recognizable by weight. I was no longer aware of dolphin presence, but I felt supported, as if in a water womb, and was able to remain focused on our work. We witnessed dramatic release, evidenced by heat and pulsing through Matt's abdomen.

After the session, Matt did not report having had any awareness of dolphin presence. However, he did say that during our work he felt deep loss and grief, yet a healing heat throughout his body. He continued to come for CST, not only because he noticed changes in his body but also because he noticed an increased awareness of his body and its reactions to stress.

His ability to integrate emotional reactions with physical sensations improved, resulting in less digestive tension and fewer migraines. He was also drawn to resume a martial arts practice, resulting in further symptom relief. The dolphins and I did not "fix" Matt, we only helped clear the way for his own natural healing to occur. The presence of the dolphins, of course, I cannot prove. Although we both felt changes, we did not generate any concrete evidence of their collaborative help in his healing. And yet, I believe in some way that dolphins continue to assist me in my work and in my life.

Sometimes I am scarcely aware of their presence. Sometimes my clients are the ones to recognize dolphin activity. I wonder if CST is not some crude relative of echolocation, the dolphin's highly sophisticated sensory skill of using vibratory waves to identify, and possibly influence, its environment. Are the dolphins interested in helping us develop our own sensory skills, so they can further share their reality with us? If so, are dolphins also interested in a collaborative healing of our Earth? Can we, in turn, offer their species safety and protection? Can human healers positively affect dolphins?

Several CST practitioners were given the unique opportunity to give something back to the dolphins. Roy Desjarlais, an instructor and therapist, was invited to work with a trio of older dolphins at an aquatic center in Florida. Two males and a female were significantly scarred. These dolphins had probably been rescued from the Navy, and they were potentially aggressive. Out of respect for their turbulent histories, they were not required to interact with the public.

Roy had been experimenting with "arcing" in the water. He found it very important to remember the concept of "neutral hands" in the water. "Neutral hands" recognizes the importance of the therapist's body as part of the healing relationship. If he projects his own fear, or expectation, or agenda through his hands, the dolphins would not choose to interact with him. If he could present "neutral hands" in the water, hands that were willing to receive information and that were capable of "listening" for vibration, the dolphins would swim under his hands, presenting the length of their spines for Roy to feel.

Roy called this "swim-by arcing." Although he describes his sensory skills as "driving a Mack truck" compared to the dolphins' echolocation as "piloting a space ship with light speed capability," he still expressed awe at the enhancement of his skills in the water with these dolphins. He described the dolphins' spines as "lighting up" in certain spots under his hands. What in a usual CST session might feel like a gentle wave to direct his hands, felt more like an electric current. The dolphins would then allow Roy to place his hands over this electric pulse, stabilizing the vertebra or section of the spine from which it emanated. Curving sideways around Roy's hands, the dolphins remained bent into a tight C-shape. Roy felt strong vibration and heat under his hands, always a sign of healing, after which he felt a letting go, or release. Taking off in a speed run around the lagoon, the dolphins could then come back for more.

The researchers at the center had been in the practice of providing traction for these dolphins by holding their flukes, or tail fins, slightly out of the water, and allowing their 500–600 pound bodies to relax. The research director suggested Roy try traction with the female. Using swim-by arcing, he felt what he identified as scarring in the soft tissue around her spinal cord just above the dorsal fin. He thought if he could hold her fluke out of the water with his hands as he had seen the trainers do, and use his feet to stabilize her spine, he could help to release her backbone and restore some mobility.

The director agreed but warned him that the dolphin would give one slight flick of the tail when she had had enough; if Roy did not let go, she would take him with her to the bottom of the lagoon, thinking he wanted to play. These dolphins had not been trained to play gently with humans. Roy was anxious, but he trusted his ability to feel for the release signal. Using his thumbs to stabilize, and his long legs for traction, Roy achieved his human-dolphin stretch, providing an action she could not achieve by herself. She softened and surrendered in his embrace, apparently enjoying the release of her spine for over a minute. When the subtle flick finally came, Roy was ready, and he released her to take off. She spun as she swam around the lagoon several times. She came back to Roy four times to have him repeat this healing. The staff later confided to Roy how unusual a display of trust his interaction was with this dolphin.

For those of us whose work is the language of trained touch and listening to the echoes of our bodies, we may be just beginning to understand the echolocation that dolphins have long mastered. Echolocation is physical. We experience it in the body as sound, tingling, and a deep sense of well-being. Can our human version of echolocation, as practiced through the listening hands of CST, compare to dolphins' skills?

Upledger is not trying to study the dolphins; he is trying to have a *relationship* with them. In fact, at the aquatic center he was adamant about not allowing sound-measuring devices in the water. To him, trying to quantify dolphin language in terms of human understanding of language symbols is not only limiting but also insulting. As he said to me, "The dolphins have told me that when we humans become more advanced we won't need to have words anymore. We'll just transfer concepts."

Could it be that our prized accomplishment, the earmark of human superiority—human language—is also what most limits our evolution? Does our obsession with naming our world and our experience with ever more specific symbols somehow obscure our greater purpose? Perhaps, in our customary beta state of brain-wave vibration, language serves us well. But could it be that the dolphins, in what may be a different predominant brain state, see *us* as the species in captivity? We are novices in their field, and we still have much to learn. Perhaps dolphins are teaching us to expand our repertoire of brain-wave experience and, thus, move beyond the limits of fear, developing greater capacity to cooperate. Maybe the next step in human evolution and healing is one of perception. Other species, such as dolphins, may have insight into a broader sensory and conscious perception. Do we have the courage to listen and learn these new skills from other animals? Only when we achieve this can we truly heal each other and, in so doing, assist in healing the planet.

The Ocean's Chalk Circle

N o one knows just when it began, or why, but gradually awareness of the change began to grow. Awareness accelerated the change. People rushed to bring the change to a full and complete conclusion. The ocean was left entirely alone, save the creatures who lived there, some of whom had once in similar fashion lost their taste for the land to return to the wet safety of the welcoming sea.

The change did not bring silence to the ocean. Its voices were now clearer and stronger than ever before. The depths reverted to a deep primordial hum free of the drone and rude interruptions of industry.

Not long after the change the orcas returned just as they had for ten thousand years. Their journey, long and at times exhausting, passed through many inlets and bays before they neared the familiar summer destination. The whales were tired but the current helped them along and refreshed their efforts. Unable to contain their excitement the young ones jumped out of the water. Everyone then rolled, tossed, and slapped their way forward. Above all they filled the whole ocean with their sounds. Call after call. A wondrous, ever changing, almost tangible curtain of sound. Then ancient traditions pulled them through the last tidal wall, into the compelling waters of the wider channel, and on to special places that promised rest and renewal.

In the seamless days and nights that followed, the orcas were aware of the change. Even though their habits, forged millions of year ago, remained unaltered, there was no longer the agitation of engines, the dangerous crowding, the confusing attentions of others. The whales had command of this marvelous world. They swam with *effortless zeal on days full of wind and white-topped waves, on mornings thick with blanketing fog, on evenings rich in golden colors. On certain nights, when the sky poured its stars and moon into a velvet sea, only the orcas, encouraged by the northern lights, dared to disturb the calm with their noisy dance. Throughout the summer, they greeted and played with the other families that came, stayed a while, left, returned, and left again.*

Like the ocean around them, they were forever moving. They traveled to the far reaches of the summer waters and explored seldom-used pathways. The younger whales, secure in the midst of their family, committed these travels to memory. From time to time, the whales passed through the thick kelp forests near the shore and felt the caress of the long sinuous strands floating in the rushing current. Sometimes the whales would break free of the surface and feel the sky against their bodies. And then there were times when all these exertions came to a stop and they moved close to each other and rested. At peace with their surroundings they became quiet. . . .

At that moment the girl awakened. Within her silent room only the thin curtains moved with the breeze from the open window and disturbed the stillness. Morning was just beginning. For a long time she lay there remembering her dream. After a while she slowly rose, went to the window, and looked out. The day was still colorless and undefined. She left the room and made her way to the shore where she sat and waited.

She heard their blows a long way off as she huddled against the cold morning. As the sun gathered strength, the girl could see their breaths rise against the light and their black shapes slip in and out of the water. They came closer, side by side, a mother and her baby not far away. The girl stood up. As the older whale passed by, her daughter moved toward the shore where the girl was standing. The whale rolled on her side and the girl could clearly see her whole body. The whale tilted her head and with a slight movement looked back at the girl. In that moment, ocean, air, and shore dissolved as the two beheld and understood each other. Then the little whale, remembering her mother and her need to breathe, swam away, leaving behind a small circle that marked her last surfacing. The girl did not move. She was aware of her own breath, quiet, soft, and small, as the circle of waves gradually widened and gently touched her still feet.

Some mornings are so achingly beautiful here, on Hanson Island where we live, it seems as if time has skipped a beat. The moment slips back into that other age when there were no humans, no adverse impacts, just wild Nature with all her adornments and profundity. We are not observers of what is now but witnesses to the commonplace of what once was. So it was on the morning of October 30, 2000. I woke up to the sounds of orcas calling me from my dream and looked out over a sky and ocean rich in predawn greys already hinting at purples, and with enough light behind and between to speak of great things to come. The morning was utterly still, the ocean a mirror of sky; an early seagull announced the day. Inside my headphones there was an eerie calm to the ocean's sound that sometimes happens as aftermath to storms, when even the snapping shrimp have been beaten into silence. Then the calls began, crystal clear, rising in my head and sweeping away whatever remnant of sleep remained from my awakening. I cannot help closing my eyes, rocking my head, and smiling. The calls are on the Parson Island hydrophone, close, there is a sudden burst of echolocation and an almost certainty the orcas are coming our way. Outside, I can hear blows in the distance, but there is a black shadow extending from the wall of Parson Island, and the whales remain hidden within it for a long time. When they are finally in clear light I can make out the unmistakable curve to the top of A32's dorsal fin. Suddenly, he launches his huge body into space and falls back to the sea; the others follow, breaching and half-breaching and spy-hopping their way past us in a spectacle as grand as any in my memory.

The next morning I am awakened at 3 A.M. The whales are in Robson Bight, an Ecological Reserve created to protect a favorite part of their habitat. They remain there for the next eleven hours. At first, the calls are loud and clear, and there is occasional echolocation as the whales grab a bite to eat. I start a digital recording in parallel to the analog tape that is running; we do this when we have the prospect of making a "good" recording. At 3:44 A.M. a small freighter hoves into view, headed south . . . boat noise on the way. For another seven minutes the whales keep calling, amidst boat noise, then they stop. Nearly half an hour later, when the engine sounds have diminished into the background, the whales start calling again. Two uninterrupted hours of the orcas calling follow, until a tugboat heads south, creating noise again. The whales stop calling and do not make another sound

I can hear for an hour and a half while first one, then another, and then another boat engine dominates the underwater soundscape. When silence returns the calls begin again. It seems the whales have not moved much if at all from where they were. Minutes later, a small boat engine intrudes and the whales stop. On this lovely morning I am irritated not just by the inconsiderate behavior of whoever was in that boat close to the whales, but also by the fact that it was inside the Ecological Reserve, a special place for the whales that people have been asked to respect.

Ever watched birds? How quiet and careful you must be, not to startle and cause them to fly away. Instinctively, you try to be invisible and small. Now consider how people approach whales and dolphins in the ocean. Usually in boats, large and small, usually with high-powered noisy engines, getting as close as possible, exclaiming with loud voices, waving arms, aiming cameras, scopes, and binoculars, shooting bullets, harpoons, and DNA darts, attaching suction cups and transmitters, jumping into the water after them, and sometimes even surrounding, capturing, and hauling them away. Then the success of the encounter is measured by how close, how many, how active, how often, how spiritual, how scientific, and even how profitable it was.

You may know the old story. A frightened little boy stands in the middle of a large chalk circle drawn on the ground. Two women who claim him as their own have been ordered to pull on him from either side until he is won . . . even if the boy is torn apart in the contest. The boy's real mother cannot bear the thought of hurting her son and so refuses. Her genuine claim is recognized and the boy is returned to her. Here we wonder if, like the boy's mother, people will be able to abandon their self-interests and stand back from the circle drawn around whales and dolphins before they and their world are irreversibly harmed.

Years ago, when orcas were "killers" and people fearful of them, they were largely avoided. Then, beginning in the mid-1960s, when "live captures" began, it became apparent that there was nothing much to fear from orcas, and, indeed, interest in them grew. This interest led directly to a boom in the marine entertainment industry that persists to the present day. Worldwide, some 133 orcas have been captured and held at aquariums and marine parks in nearly a dozen countries. The principal capture sites have been in the Pacific Northwest, Iceland, and Japan. Today, as few as 24 of the wild-caught orcas survive, and captures are prohibited in most of the former locations. The captive industry increasingly relies on in-house breeding programs and on between-facility "loans" to sustain sufficient numbers of orca performers to please crowds, though captors are now being welcomed in Russian waters. For the industry the captures have been a boon; for the orcas it is certainly a different story, though we will never know that entire cost. So little is known about the orcas that inhabit Icelandic and Japanese waters that we can only guess at impacts. What we do know is that the "southern resident" community of Pacific Northwest orcas lost virtually an entire generation to captures in the 1960s and 1970s; that "A clan" of the "northern resident" community of British Columbia lost an entire matrilineal family group and many members of several others; that numerous unborn were lost to future generations; that many potential mothers were lost also; that the lives of captives are a dismal shadow of the world they came from; that captives have died too often and too young; that whatever spin is put on the "educational" or "scientific" benefits of keeping orcas in captivity, they pale into insignificance compared to what we see of orca life in the ocean.

One August morning the whales were far from our hydrophones, so we busied ourselves with other matters. Visitors came by. We had tea and talked. Then we heard the calls. The whales were returning. Soon, they turned the corner and came into view. By now we knew there were many families present. The orcas were spread out across the Pass in small groups of twos and threes. The sun was low in the sky and the colors of the sunny day were beginning to deepen. Although the whales were in a hurry, one small family paused opposite us and drifted at the surface. The oldest son, Cracroft, turned back and seemed to be waiting. Then we saw Sharky and her two young ones heading slowly toward him. When they finally reached him, Cracroft turned and followed them in the direction of the other whales who had disappeared into Blackfish Sound. It was altogether a lovely afternoon moment. But before we had time to reflect on what had happened, the whales surprised us again, returning from Blackfish Sound almost as soon as they had gone out of sight. This time they were so closely bunched together it seemed impossible they had room to breathe. Forty whales formed a tight, rolling ball, with heads, fins, and flukes poking through, heading back through the Pass. What had excited all this? And where was Tsitika's family, the only group not there? We kept our eyes focused on the moving orca ball. We could hear the exclamations of the visitors on the deck of their boat, moored just beyond the kelp in front of the Lab. By now the sun was even further down and the afternoon light played out in rich hues on the calm sea. Then suddenly, literally feet from shore in front of the Lab and facing us, Tsitika's family rose and gently broke the surface. Six whales breathing as one—mother, three huge sons, and two daughters, so tightly packed together they must have been touching. There had been no hint they were near. Twenty witnesses were too stunned to move. Everyone stared as the whales sank from sight, then surfaced and resurfaced, in the same place, again and again and again. Eventually, amidst an uncanny human silence, the whales sank one last time into the dark undisturbed waters and swam away to join the rest of their kind.

Though captive displays did much to dispel the basic fear-provoking myth, they also created a set of expectations about seeing orcas up close and personal. These days, the first impulse of many people who see orcas when they are out in a boat—perhaps being there primarily to enjoy the day or do a spot of fishing—is to want to get close to the whales, to see them full-frame, get the "hit" and take photos and videos back home. There is of course no harm in people wanting any of these things, but in getting them they too often intrude on whatever it was the whales were doing beforehand. Because resting orcas are easiest to approach—usually swimming slowly together in a group that presents a perfect photo opportunity—such intrusions have become commonplace. Rest is obviously a vital aspect of orcas' lives, so it is easy enough to see the potential for cumulative negative impacts. Quite possibly, an occasional disturbance of rest creates no problem at all for the whales, but when it is repeated time and again, it must. The same logic is true for other vital orca activities, such as feeding and socializing. A few disturbances during the course of a day probably do not matter that much, but when they become constant in the normal flow of living, they probably matter significantly.

When people first began to pay close attention to the orcas of Johnstone Strait, one of the common sights was of "resting lines" that would sometimes stretch literally from side to side across the three-kilometer-wide strait. These consisted of family groups of resting orcas, members traveling closely side by side in a horizontal formation, with gaps between families. The whales would sometimes spend hours slowly moving along together in this wonderful display of coordinated resting. It seemed to us at the time that these resting lines had a social significance as well as a biological one. Not much later, when people started seeking out "experiences" with orcas, the resting lines all but disappeared from the scene, the reason probably being the ease with which approaches by boat could be made, and the frequency with which that happened. This is not to say that the orcas stopped resting, but rather that the elegant manner in which they went about it together changed. Nowadays members of orca families usually rest separately in their own groups, rather than collectively with other families, though in the past few years, perhaps coupled with a growing understanding of the whales' needs and a heightened sensitivity to them, the "resting line" phenomenon has occasionally reappeared. If resting lines do become a common part of the summer orca scene again, it will be because people have learned enough about the whales' needs to respect them and have modified their own behavior accordingly.

Part of the whales' dilemma is that it is no longer difficult for people to enter their space. For humans, the idea of being in wild places is no longer imposing. It is still true that every so often Nature catches an unprepared visitor's attention with an unsuspected storm, with wandering bears or cougars that scare, with tides or currents that overwhelm. But for the most part summer rolls on with surprisingly few incidents, and visitors have come to view the most popular places to see orcas as safe playgrounds. As a result, behavior is no longer naturally influenced, and this change is reflected in increasingly casual attitudes toward wilderness.

One day in 1999, we heard a report of a possible birth in a particular group and wanted to find out if it were true. So we ventured out into Johnstone Strait in our boat, and for a while watched the whales in the distance as they traveled along the Vancouver Island shoreline. Anticipating that they would pass near Izumi Rock we headed that way, disembarked, and waited. Izumi rises out of Johnstone Strait about 100 meters from Vancouver Island. It is always exposed, and at low tide it is prickled with barnacles and slippery with seaweed. The water around the rock is deep enough for boats to approach safely and for whales to pass on the inside, between it and the island. We had seen kayakers come off Kaikash Beach (a popular camping spot about three kilometers west of the rock) and paddle energetically directly toward the whales, who were then fairly close to shore. But as the kayakers paddled out, the whales altered their direction and speed. Gradually, they moved farther off shore and steadily increased the distance between themselves and the kayakers. The shift was subtle. The kayakers had to work very hard and never got close, but they persisted and followed the whales down to Izumi.

When the whales passed Izumi, they were traveling close together. We balanced on the rock, peering intently through binoculars, a spotting scope, and a video-camera viewfinder. The sun shone brightly, creating highlights on the gleaming black fins as they rolled into the ocean after one another. We had a clear view and were sure there was no new baby. But we could not be disappointed in the face of such power and beauty. It was a marvelous moment, and we were still intensely preoccupied with it, following the whales as they headed away, when suddenly a voice intruded, from below us, "Do you know what the weather forecast is?" It took us a while to juggle realities and get onto the same plane so we could stutter out a reply as the kayaker paddled on her way. Possibly our answer was a bit abrupt. It was not that the request was unreasonable. The kayaker had probably seen our boat from a long way off and had been thinking about asking her question for some time. What made it strange was the feeling that we might have been anywhere, answering a question for a casual passer-by on some downtown street corner. Our unusual location on a rock in the middle of wild Johnstone Strait, our single-minded focus on the whales, caused no curiosity or comment. It was just part of the scene on another Johnstone Strait summer's day.

A fundamental problem for animals that inhabit the oceans is that we who live on land find it very difficult to relate to what happens in their world. "Out of sight, out of mind" could not be truer of attitudes toward the oceans, from the relentless profit-driven toxic outpourings of worldwide industry to the casual dumping of untreated human waste by city authorities who know better, to the purposeful pumping of bilges at sea before ocean-going vessels reach port, to the hideous assault that the world's principal military powers are presently planning and testing, soon to begin deploying—the so-called "Low Frequency Active Sonar" (LFAS) systems that have been devised as a means of detecting "silent" submarines at great distances. The idea behind LFAS is simple enough: to bounce sound waves off an object at a distance and figure out what it is. The problem is the distances are huge, sources of noise are many and varied, and the targets are tiny. So you have to use very loud sounds to get anything meaningful back. Loud is a relative term that anyone who has attended a rock concert understands, but in this case "loud" is literally unbelievable and inconceivable in its intensity and impact. LFAS sounds harbor lethal energy. They cause pain, injure, maim, and kill—and this is especially so for acoustically sensitive whales and dolphins. Despite clear evidence from cetacean strandings, injuries, and deaths that coincided with LFAS tests in the Mediterranean in 1998 and naval maneuvers near the Bahamas in 2000, the rush toward deployment continues. Meant to improve "national security," LFAS amounts to nothing less that waging war on whales. Soon, if LFAS is not stopped, the world's oceans will become polluted by sound pulses at levels that will make life difficult to impossible for their cetacean inhabitants.

Sadly, science and scientists are partly to blame for this sorry state of affairs. One might think it a simple matter to string together the chain of logic. Filling the ocean with unbearable noise is obviously a bad thing to do to acoustically sensitive creatures. Yet time and again we find huge efforts and resources being put into "scientific" evaluation of the impacts of some of the grossest manipulations of the natural world imaginable. LFAS stands out as a peak in this mad zone, but before that there was Acoustic Thermography of Ocean Climate (ATOC), and before that "ship shock." We should not speak of "before" as though these things were a past matter, because they are still ongoing horror stories. Ship-shock tests—seeing how well ship hulls stand up to high explosives—are a straight-on military assault on the ocean. The problem they create is obvious to anyone who has seen a stick of dynamite tossed into the water; it is about the easiest way to kill fish. The ATOC experiment comes with heavy consequences for living things. Physicists and the U.S. government (read: military) have managed to put together a scientific team with sufficient credentials and clout to cloud the basic issue beyond any capability the public might have to sift through facts and arrive at conclusions. Tell the public often enough there is no risk and apparently there is none—as long as there are nodding scientists standing beside you. In the case of LFAS deployment, it should be a simple act of humanity for the scientific community to stand up en masse and say NO. It is with the deepest regret that we note that Science, seeker of truth, has turned out in too many instances these days to be more of a malleable hireling than a harbinger of hope.

Central to the malaise are nationalists who put science to work promoting morally bankrupt policies of governments seeking to maintain, by whatever pretext, "rights" and "traditions" the world at large has rejected or moved on from. Outright zealotry accompanies Japan's stubborn insistence on killing whales. The blood sports of the Faeroes come from a very dark place in the human genome. Both are defended by scientists who insist there is no problem, even that killing whales can solve the problem of dwindling fisheries because whales eat fish. Were the consequences of such thinking not so serious, one could be amused. But the fate of whales is no laughing matter. Indeed, it is our view that the fate of whales has much to do with the fate of planet Earth. Whales became an icon for caring in 1972 at the first United Nations conference on the environment. Held in Stockholm, it called for a worldwide moratorium on commercial whaling, based on incontrovertible scientific evidence of dwindling populations, as a symbol of the struggle to "save" our planet. Ten years later the moratorium was agreed to by the International Whaling Commission (IWC), the body charged with the responsibility for managing whaling and protecting whales. The victory was a virtual miracle because it required a three-fourths-majority vote. Moreover, the moratorium was indefinite; it held no time limit. One might be excused for thinking it would last forever, that whales were truly "saved." Not so. A key provision of the treaty under which the IWC was established required it to further "the orderly development of the whaling industry." There is little to no question that "orderly development" should now focus on economic activities based on living whales, such as whale-watching, not on reopening the killing grounds. Good reason for maintaining the moratorium, one might think. Yet today the IWC stands on the brink of voting to overturn the moratorium and to sanction killing whales for profit again. Unbelievable, yes, but true.

It is hard to know where on the circle of influence to place the public. We (the masses) should perhaps be at the center, but it is hard to see us even on the periphery. Public consultation about policy issues—often mandated by law—too often amounts to lip service and pretext. Public-interest advocates are given opportunities, even asked for input, but they have to create a literal storm of comment and protest before their voices are heard. Because apathy rules the public day, those with access to power, money, and media manage to control information, perceptions, and attitudes on a vast scale, with little to no awareness on the part of those whose strings are being pulled. How else to explain what is happening to whales these days? It is difficult to understand why such obvious and widespread abuses—killing, capturing, "tagging," chasing—are tolerated today. Why doesn't the public rise up in arms at the spectacle of beautiful, innocent dolphins and whales being driven ashore and butchered? Why does the spectacle of a hundred boats crowding around half a dozen whales, giving them no room to move, continue to be tolerated? Why isn't a single questioning voice raised in a public forum when a "scientific" project that entails shooting radio "tags" into orcas' bodies is described? The answer eludes us, but like the key to an unsolved mystery, it is there to be found. We take as a clue a perceptive comment made in a recent radio discussion about perpetuation of patterns of abuse in human communities, namely that with love comes responsibility. One can love, but still neglect. Together, love and responsibility create an irresistible positive force.

In the short term, at least, some problems can be managed as we make progress in our understanding of the needs of cetaceans. A mix of learning, empathy, and recognition of our care-taking role will take us a long way as we pursue our fascination with whales. For the whales, should we be impelled to take the steps back from the circle that are needed, the mix may be enough to give them a life.

"Necessary Kindness": An Interview with Jean-Michel Cousteau

Jean-Michel Cousteau is an explorer, environmentalist, educator, and film producer, communicating to people of all nations and generations his love and concern for our water planet. He is an impassioned, eloquent spokesman for the environment, continuing his pioneering work through Ocean Futures Society.

The son of ocean explorer Jacques Cousteau, Jean-Michel spent his life with his family exploring the oceans aboard the research vessels Calypso *and* Alcyone. *He has produced over seventy films, interactive media, and a syndicated column. He has been awarded the prestigious Peabody Award, the Ace Award, the 7 d'Or, and the Emmy. In addition, in 1998 he was awarded the Environmental Hero Award by the U.S. government. During the 2002 Winter Olympic Games, he represented the Environment as he carried the Olympic flag.*

I first met Jean-Michel in 1998 in Baja, Mexico, where he was filming a documentary and I was working on a book for National Geographic, both centering on the gray whale. Since then, we have dialogued together on other issues, including the Makah gray whale hunt, Keiko, and dolphin conservation. In this fall 2000 interview, Jean-Michel shares with us how his devotion to the world's oceans inspires his life and work. — BP

In Baja, you spoke about the "necessary kindness" our species needs in our kinship with other species. How do humans develop this empathy? Let's start with a story. How did you come to know dolphins and what was your first bond?

My strong personal experience with dolphins came when I was fifteen. My father [Jacques Cousteau] was director of the Oceanographic Institute in Monaco. He wanted marine mammals, especially tursiups [bottlenose dolphins], as part of their aquarium. So they sailed out to capture dolphins. Jacques Cousteau and Albert Falceau had designed pincers, like those tongs used to grab sugar cubes. You stand on the bow of the boat, open the pincers, and throw it onto the tail of dolphin. Today you go to jail if you do this. This was a long time ago. They captured ten or twelve dolphins over time, and every time one was captured he or she was taken into an Olympic-sized swimming pool in Monaco. We were called upon to literally get hold of dolphins and "walk" them around the pool to show the dolphins the limitations and walls of the pool. It would take two to four days and then we could put them in a smaller pool, like the size of an aquarium pool. We realized that there were three kinds of dolphins: those dolphins who refused captivity and quickly died; other submissive dolphins who completely accepted captivity and gave up; and those dolphins who could not make up their mind.

I ended up being in charge of a captured dolphin in the smaller pool. My job was to spend time with him and see how he was doing. At the time, I was going to high school. I'd first go to see my dolphin, come home, see the dolphin, do homework, see the dolphin. I knew that this dolphin was not doing well. Something was fundamentally wrong. I kept calling the vet, who would come take his temperature and say, "No, he's fine. No problem." This was completely a physiological reading. One day I went back and the dolphin was dead. He just raced from one end of the pool to the other, blew his brain out against the wall—he committed suicide. That death has stayed with me all along.

How did this early death change your ideas of captive dolphins?

I have a completely different view now of these animals. Keiko, the killer whale made famous in the movie *Free Willy* and now released back into the wild, is reminding me of this every day. I consider orcas, the largest dolphins, as being to the ocean what humans are to the land—with a lot less vicious aspects than we have. They are a dominant predator and nobody kills them. That's what we do, too. We even go and kill orcas as well. So dolphins are a very important part of my life. They have deeply affected me.

I've always been against captive dolphins. It's slavery. It's putting these

animals in jail. It's an acoustic jail and that's why we humans cannot understand or relate to it. This enslavement is difficult to convey. Orcas live in a
world of sound. Suddenly you create an echo chamber of concrete aquarium
pool walls, in which the dolphins are completely disoriented. Ultimately they
shut up or go nuts. We have no right to do this to dolphins for the sake of
our commercial pleasure. My exceptions are in a situation such as an injured
bald eagle, emblem of the United States. You are not going to shoot or kill
the bald eagle. If the bald eagle has a broken wing and doesn't fly, you take
care of the bird. You use him as an educational animal ambassador in schools.
Give him a chance to keep his pride and nurture the interest of young people.

There are some dolphins who, if you release them or if they've been
maimed—whether from the United States Navy doing sonar experiments
that negatively affect their hearing systems or injury from a boat accident—
that animal will never make it in the wild. So you take care of these animals.
You give them a place, a closed bay, where they can be. People can look. But
to have them jump through fire hoops—this is antieducational.

To prove dolphin intelligence, we expect to make them think and reason
and talk as we do. Their world is a different world. I'm totally opposed to
using these animals for selfish pleasure. This teaches our children the wrong
thing. We wouldn't do this with lions. Lions are creatures that can be nasty.
So you don't put lions in captivity and then go and play with them. I think
to do that with dolphins is just as cruel. They will tell you it's good for handicapped people. That's all an excuse. I could take handicapped people out
into the ocean and they can have the trip of their lives.

What is happening with the captive dolphin industry now?

That's a major problem now with so many captive animals. For example, in
Mexico City, two beluga whales are being bought from Russia and now exist in a tank under a roller coaster, between noise and vibration. I went to
see those two beluga whales. One of them spent twenty minutes banging
his head against the wall.

In your introduction to the Dolphins *IMAX–National Geographic book, you write
that your father once told you, "People protect what they love." What can ordinary
people do to protect the animals and environment that they love?*

We can reinforce the laws and regulations. It works in industrial countries.
In Third World countries, the only way is education. Unfortunately, money

talks and often we get outvoted because of money. If we can reach young people, then it's ingrained and when they become decision-makers they can do better by dolphins and other marine mammals.

Since Keiko, the killer whale of Free Willy *fame, was released from a Mexico City tank, he has been in rehabilitation in Oregon and then in Iceland, his native home. Can you tell us about Keiko's progress toward eventual release into the wild?*

Keiko has ventured out in the open ocean and has had many encounters with other whales. Each time other whales either departed or Keiko departed. One very primitive way of looking at this is that Keiko has not yet found his right pod. There are many pods. We haven't had Keiko meet all of them. We've done a lot of recording, and we know that Keiko speaks the language or vocalizations of North Atlantic orcas. We've done a lot of photo ID studies of many whales. We have to completely review his position regarding his bent dorsal fin—which we thought was due to so much time in Mexico in high altitude with his dorsal out of water for hours every day. We've found a pod up in Iceland which may be Keiko's. It has two male orcas with bent dorsal fins. We've filmed them. So the bent dorsal may have to do with genetics, not captivity. That's another big question.

Keiko is extremely healthy. He can both fish and feed on his own and come back to us to be fed. We always provide him with fish, but we don't give him food by hand into his mouth. We throw it in the ocean. So it looks like it just happens to be there. We are still in the spirit of disconnecting with him. This detachment is hard for him; it's even harder for us.

Do you have any insights on why people are so fascinated by and fond of dolphins?

Let's not forget where they come from. They're like dogs. These things never go away. We were evolutionary hunters and gatherers. We were not hunting the dolphin dog. They were part of our family, our tribal system, even before the human species appeared. They made their way into the ocean fifty to sixty million years ago. But there is something that connects us to dogs. Look at man and wolves. We've always been together. I'm talking about a time when the human species did not exist. That's when the dolphins were on land. We were not around. In evolutionary terms, we're new kids on the block. But we have no elders to tell us what to do and not to do. We humans screw up all the time.

What should we be doing to help conserve and protect dolphins and other marine mammals in the twenty-first century?

We need to understand them better. We don't want to teach them English. We need to understand their communication system. With all respect for Dr. John Lilly, I think he went about it the wrong way. We need to spend more time at sea to understand how dolphins function, work, and behave, and exactly what is their social structure. We cannot do this in a bay pen and in a tank. We have to be out there in the sea with them. That's true for all cetaceans. Because they don't spend time on land. It's time for us to go out there and not for them to come to our world.

All captive studies are limited in their scope, because the dolphins don't behave or communicate normally. They don't eat normally. What are we learning besides physiology? You don't take a Papuan from New Guinea and bring them into a little park in New York City to study them. You go to Papua New Guinea.

So both as scientist and as conservationist, you believe that we need to redefine our proper study of cetaceans. Tell us about your work now with Ocean Futures.

Ocean Futures is a voice for the ocean. If you protect the ocean, you protect yourself. We have five critical issues:

- *Water quality.* Today when you buy water, you pay between five and twenty times the price of gasoline.

- *Coastal habitats.* These are the nurseries of the ocean. Over 50 percent of the world's mangroves have been cut down.

- *Coral reefs.* These reefs are dangerously in jeopardy now. At a recent environmental conference, I made a statement signed by many scientists that is frightening. It says that, in the next ten years, we may lose as much as 40 percent of our coral reefs, and, in the next twenty years, we could lose another 20 percent. Coral reefs are the rain forests of the ocean.

- *Fisheries aquaculture.* Fisheries are being depleted at an alarming rate, with 62 percent of fish stocks urgently in need of better management. If we can create action partnerships with businesses, industries, and international governments, we could find cooperative solutions for a "Blue Revolution" in this twenty-first century.

- *Marine Mammals.* Why marine mammals? They are the top of the food chain and the great indicators of the health of the oceans. We've been studying killer whales in the Pacific Northwest since the 1950s. The population isn't increasing because they're being poisoned in their environment, by the toxins in their food, the heavy metals and PCBs in their habitat. Now orcas aren't multiplying the way they used to; their young are dying at an alarming rate. The cetaceans tell us all the things that humans are doing wrong. Orcas spend more energy to get the same amount of food or less.

Imagine ahead to the year 2050. What do you see, as a visionary, in the best of all our possible worlds?

We will have stabilized our population growth. We will have found educational ways to help people understand what it is to manage finite resources in a sustainable way. In order to do this, we have to understand what is there, how it works, how to assess populations, and thus figure out what is the interest on the capital that we can use. Right now we are gobbling up the capital of our natural resources. It's like a bank account. Nature works for free, that's the big difference.

I hope it won't be the orgy of greed that we are experiencing now. Third World countries now want to capture more animals. But perhaps they can learn faster than we did.

I am an optimist. I believe we are going to make it. Humans can learn fast—if we have to do so. Most of all, we need to trust the new generations.

As we go to press, Keiko has made remarkable progress toward becoming a wild whale. In the third year of reintroduction efforts from his open-water pen in Iceland, in June 2002 Keiko spent three weeks in the close company of orca pods. Over the next month he traveled more than a thousand miles eastward to the Norwegian coast. During that time, dive-time data, coupled with photos and girth measurements showing no weight loss, have led his veterinarian to the firm conclusion that Keiko has proved his ability to successfully forage and sustain himself in the wild. After arriving in Norwegian waters, Keiko entered a fjord and reverted to his curiosity toward boats and people before regulations were implemented. But now that swimming with, feeding, and closely approaching Keiko is prohibited, Keiko is being readied for further efforts to connect him to wild whales. Keiko project managers intend to continue to allow Keiko to make the choice about returning to wild orca populations. This is exciting news and holds out great hope that Keiko may become the first orca to be successfully returned to the wild. — BP

Advice to a Prophet

When you come, as you soon must, to the streets of our city,
Mad-eyed from stating the obvious,
Not proclaiming our fall but begging us
In God's name to have self-pity,

Spare us all word of the weapons, their force and range,
The long numbers that rocket the mind;
Our slow, unreckoning hearts will be left behind,
Unable to fear what is too strange.

Nor shall you scare us with talk of the death of the race.
How should we dream of this place without us?—
The sun mere fire, the leaves untroubled about us,
A stone look on the stone's face?

Speak of the world's own change. Though we cannot conceive
Of an undreamt thing, we know to our cost
How the dreamt cloud crumbles, the vines are blackened by frost,
How the view alters. We could believe,

If you told us so, that the white-tailed deer will slip
Into perfect shade, grown perfectly shy,
The lark avoid the reaches of our eye,
The jack-pine lose its knuckled grip

On the cold ledge, and every torrent burn
As Xanthus once, its gliding trout
Stunned in a twinkling. What should we be without
The dolphin's arc, the dove's return,

These things in which we have seen ourselves and spoken?
Ask us, prophet, how we shall call
Our natures forth when that live tongue is all
Dispelled, that glass obscured or broken

In which we have said the rose of our love and the clean
Horse of our courage, in which beheld
The singing locust of the soul unshelled,
And all we mean or wish to mean.

Ask us, ask us whether with the worldless rose
Our hearts shall fail us; come demanding
Whether there shall be lofty or long standing
When the bronze annals of the oak-tree close.

Interacting
with Captive Dolphins

Toni Frohoff

In 1860, P. T. Barnum captured two beluga whales for public display in New York. Since then, cetaceans have been captured, transported, and maintained in many countries around the world. Accurate numbers of captive dolphins or existing facilities are not possible to obtain because many facilities are not licensed, most countries do not require inventories, and the numbers are constantly changing. However, approximately 200 captive dolphin exhibits are currently *known* to exist worldwide in approximately sixty countries. The controversy over the ethics of captive dolphin public display facilities has increased greatly over the past few decades. (Rose, in this volume, provides data on captive dolphins, including statistics on dolphin longevity; Bekoff, in this volume, discusses ethics.)

Captive dolphins are sought after as sources of entertainment and recreation and, to a much lesser degree, for therapeutic, research, and educational purposes. An increasing number of facilities offer interactive programs with cetaceans, such as petting/feeding programs and "swim" programs to the public for a fee. The few studies that have been conducted on captive interactive programs all observed stress-related behaviors in dolphins that were related to potentially long-term negative physiological states.[1] And the impacts of capturing dolphins from the wild can negatively impact not only the individuals captured but also the populations from which they are taken. The educational benefits of these programs and their safety for people are also highly questionable. Regardless, these programs not only continue but are also expanding at an astonishing rate internationally, especially in Asia and the tropics, largely because of tourism-related revenues. In one recent year in the Caribbean alone, facilities have opened in Anguilla, Antigua, Jamaica, and Tortola, and others are in development in the Cayman Islands, the Dominican Republic, St. Lucia, Curaçao, and Dominica.

Laws governing the capture and maintenance of dolphins in captivity vary widely around the world, from being nonexistent to providing detailed regulations. The degree to which regulations are enforced is also highly variable. Of the countries that have implemented laws to protect captive dolphins, most (including the United States, at the time of this writing) have no special provisions that have been enforced to protect dolphins used in swim programs or petting/feeding pools. (Because these laws fluctuate so widely and frequently, readers seeking more details are encouraged to visit the Whale and Dolphin Conservation Society's Web site at www.wdcs.org.)

DOLPHIN SWIM PROGRAMS
IN THE UNITED STATES

Currently, in the United States, approximately twenty facilities offer some variation of swim programs with captive dolphins (including programs in which participants can "wade" with the dolphins). These facilities range from sea pens in tropical waters (such as those in the Florida Keys) to concrete tanks (such as one at a hotel in the heart of Las Vegas). In the United States, the Department of Commerce's National Marine Fisheries Service (NMFS) lost its regulatory authority over swim programs in 1994; the U.S. Department of Agriculture's Animal and Plant Health Inspection Service (APHIS) now has sole jurisdiction over them.

In 1998, APHIS adopted new regulations regarding swim programs (after allowing the programs to operate for over four years without specific regulation). However, as of April 1999 (approximately four years before this book went to press), enforcement of these regulations was suspended. Therefore dolphins used in swim programs in the United States have still not been offered any more protection than those who are not used in these programs. Nor have these programs been explicitly authorized. (APHIS regulations can be viewed at www.aphis.usda.gov/ac by looking at section Part 3, Standard, under Publications.)

PETTING POOLS IN THE UNITED STATES

This section adapted from: C. Williamson, "What Lies Beneath the Dolphin's Smile? The Case Against Dolphin Feeding and Petting Pools," Whale and Dolphin Conservation Magazine 24 (Spring 2001): 20–21.

The Whale and Dolphin Conservation Society (WDCS) undertook the first study to examine captive dolphin feeding and petting pools. These pools provide the public with the opportunity to touch and feed captive dolphins from poolside. Currently, there are no special regulations governing the care and protection of dolphins in petting and feeding pools. Current humane handling and care regulations set under the Animal Welfare Act that govern the treatment of marine mammals in captivity do not address the unique features of petting pools or the special conditions that confront dolphins in these environments.

The WDCS conducted their observations at four Sea World facilities in the United States (subsequently, one of these facilities was sold to another company). According to the WDCS, the study provided indisputable evidence that physical interaction between captive dolphins and the general public can place both parties at a significant risk of harm. It concluded that the intense and poorly regulated physical interaction between inexperienced members of the public and dolphins in crowded petting pools poses unacceptable physical risks to both humans and dolphins.

During the main tourist season, visitors to Sea World often have uninterrupted access to petting pool dolphins for as many hours as the park is open—up to fourteen per day. At the busiest times, several hundred visitors can crowd around the pool's edge, frequently splashing water and banging on the sides of the pool using their hands, coins, and other hard objects. Additional noise is produced by music played over loudspeakers, staff talking into microphones, screams from children, and noise from music events and firework displays within the park. So each and every day the dolphins are subjected to a constant source of often intense stress and disturbance from which they cannot escape.

The investigation recorded visitors commonly teasing dolphins with fish and touching the dolphins inappropriately (for example, near or on the blowhole). Several incidents of risks to people were observed, including bites, head butts, and trapped hands. In one case, a boy was observed being hit in the face by a dolphin's head. Other serious problems included objects such as sunglasses, hair accessories, and human food items and contaminated fish either being dropped into the pool or intentionally offered to the dolphins. The dangers of swallowing foreign objects are confirmed by numerous examples of mortality among captive dolphins that have swallowed objects introduced into their pool. Obesity and nutritional concerns in dolphins were also observed.

In 1999, the WDCS submitted a report to the U.S. Department of Agriculture calling for a full investigation into the problems identified at petting pools in this study and proposing that, in the absence of adequate action, such pools be closed down. Follow-up visits by the WDCS through 2001 demonstrated that the welfare of both dolphins and visitors continues to be placed at risk at these facilities. A copy of this report and updates on this issue can be obtained by contacting the WDCS at www.wdcs.org.

NOTE

1. T. G. Frohoff, "Behavior of Captive Dolphins (*Tursiops truncatus*) and Humans During Controlled In-Water Interactions" (Master's thesis, Texas A&M University,

College Station, Texas, 1993); A. Samuels and T. Spradlin, "Quantitative Behavioral Study of Bottlenose Dolphins in Swim-with-the-Dolphin Programs in the United States," *Marine Mammal Science* 11 (1995): 520–44; B. Maas, S. Fisher, C. Williamson, C. Stark, and M. Simmonds, "Behavioral Patterns Exhibited by Captive Dolphins at Feeding/Petting Facilities," in *Abstracts of the 13th Biennial Conference on the Biology of Marine Mammals* (Maui, Hawaii, 1999).

Interacting with Dolphins in the Wild: Science, Policy, and Guidelines

Toni Frohoff

The popularity of programs offering the public opportunities to interact (in water or from boats) with free-ranging cetaceans at close range has increased substantially over the past two decades worldwide, largely because of the enormous expansion of formalized and commercial ventures. These interactions sometimes involve the provisioning of food to the animals. Viewing dolphins and whales in the wild can be done responsibly. However, research on and management of these programs have lagged dramatically behind their expansion. Consequently, individuals as well as populations of dolphins are often subjected to various forms of stressors, injuries, and even mortalities as a result of "sociable" human contact, though generally unintentionally.

SCIENCE

In 2000, the International Whaling Commission formally addressed in-water interactions with free-ranging dolphins for the first time. For this meeting, an independent review of the literature on this subject was conducted and revealed the following for odontocetes (toothed dolphins and whales).[1]

- Even the most well-intentioned sociable human interactions with odontocetes are accompanied by an unknown number of unpredictable impacts and risks to the animals, some of which may be cumulative, long-term, and life-threatening to individuals as well as to populations.

- The impacts of swimmers, even in the absence of boats, on odontocetes may be serious under a wide variety of conditions.

- Odontocetes exhibiting the highest degree of contact with humans are generally at the greatest risk of injury, illness, or death from humans.

- Various aspects of human activity and animal characteristics are important factors in evaluating the potential for impact to animals. The observed impact of numerous human swimmers was reportedly minimal in some situations, whereas the impact of a single swimmer in other situations was severe enough to result in repeated disruption of rest and other important activities.

- Because of the many difficulties inherent in observing behavioral impacts to odontocetes, determination of no or minimal impact to cetaceans should not be made based on the absence of perceived short-term responses alone.

- Management of interactions with odontocetes, when implemented, is often wholly inadequate and lacking appropriate research, monitoring, education, and enforcement proportionate to the extent and intensity of the interactions.

This review concluded that a precautionary approach to managing these activities is warranted so that the burden of proof lies with those who want to interact with the animals rather than those who want to protect them.

POLICY

Regulatory protection of cetaceans ranges widely from nonexistent to stringent around the world. The U.S. Marine Mammal Protection Act (see below) has among the strictest provisions for protecting dolphins from human activity (see http://www.nmfs.noaa.gov/prot_res/laws/MMPA/MMPA.html). The Action Plan for Australian Cetaceans also offers a high degree of protection to dolphins and whales (see http://www.ea.gov.au/coasts/species/cetaceans/index.html). The New Zealand Marine Mammals Protection Act and regulations also offer detailed protection to marine mammals and is somewhat unique in offering permits for in-water interactions with dolphins (see http://rangi.knowledge-basket.co.nz/gpacts/public/text/1978/an/080.html).

However, regulations alone are frequently not sufficient for protecting dolphins. In New Zealand's Milford Sound, for example, a population of dolphins is being threatened by tourist boats that come to see the dolphins. Biologist David Lusseau recently found that dolphins are being injured and killed by the boats, and up to 7 percent of them bear visible scars from boat collisions. He documented that two out of three encounters between boats and the dolphins involved the boaters violating New Zealand's Marine Mammals Protection Act.[2]

U.S. LAWS

The following section was adapted from Trevor R. Spradlin, Jeannie K. Drevenak, Ann D. Terbush, and Eugene T. Nitta, "Interactions Between the Public and Wild Dolphins in the United States" (presented at the Wild Dolphin Swim Program Workshop held in conjunction with the 13th Biennial Conference on the Biology of Marine Mammals, Maui, Hawaii, November 28, 1999).

The National Marine Fisheries Service (NMFS) is the U.S. federal agency responsible for administering the Marine Mammal Protection Act of 1972 (MMPA; 16 U.S.C. *et seq.*) for the conservation and management of all cetaceans (whales, dolphins, and porpoises) and most pinnipeds (seals and sea lions). The MMPA and its implementing regulations promulgated by the NMFS prohibit the harassment and feeding of wild marine mammals. The MMPA carries both civil and criminal penalties for violations: the maximum civil penalty is $10,000, and the maximum criminal penalty is $20,000 and one year in jail.

The NMFS is concerned that recreational interactions between the public and wild dolphins—specifically swimming and feeding activities—are disturbing the dolphins ("harassment") and that the health and welfare of the animals are at risk. Several of the "swim-with-the-dolphin" (SWTD) programs in the southeastern United States appear to be enabled by either feeding the dolphins or habituating them to human presence over time.[3] Feeding has been determined to be harmful to the health and welfare of wild marine mammals and was specifically prohibited by the MMPA regulations promulgated in 1991.[4] It is illegal to feed or attempt to feed a wild marine mammal either food or nonfood items. The SWTD programs in Hawaii encroach on the sensitive habitat areas that the dolphins use for resting and sheltering.

In addition to the biological concerns, there are several legal issues regarding the MMPA and its regulations that need to be considered. Under the MMPA, it is illegal to "take" any marine mammal. The term "take" is defined as:

> to harass, hunt, capture, or kill, or attempt to harass, hunt, capture, or kill any marine mammal.

The prohibition on "taking" applies to all marine mammals within U.S. jurisdiction, and there are limited exceptions to the prohibition for permitted "takes" pursuant to specific activities such as: scientific research, public display, enhancement of a species or stock, commercial/educational photography, and incidental take in commercial fisheries.

In 1991, the NMFS amended the definition of "take" in the MMPA regulations to include a prohibition on the negligent or intentional operation of an aircraft or vessel, or the doing of any other negligent or intentional act which results in disturbing or molesting a marine mammal; and feeding or attempting to feed a marine mammal in the wild. (50 CFR 216.3)

In 1994, the U.S. Congress amended the MMPA and defined the term "harassment" as "any act of pursuit, torment, or annoyance" which:

1. has the potential to injure a marine mammal or marine mammal stock in the wild (Level A harassment), or

2. has the potential to disturb a marine mammal or marine mammal stock in the wild by causing disruption of behavioral patterns, including, but not limited to, migration, breathing, nursing, breeding, feeding, or sheltering (Level B harassment).

Thus, "harassment" can be an act of *pursuit* that has the *potential to disturb behavior*. The NMFS is concerned that SWTD activities in the wild risk causing harassment to the dolphins because, by their nature, they pursue interactions with wild dolphins that can disrupt the animals' natural behavior.

The NMFS recognizes that there are situations in which wild dolphins will approach people on their own accord, either out of curiosity, to ride the bow wave, or to surf the stern wake of a vessel under way. If wild dolphins approach a vessel, the NMFS recommends that the vessel maintain its course and avoid abrupt changes in direction or speed to avoid running over or injuring the animals. If wild dolphins approach swimmers or divers, the NMFS recommends that the people avoid abrupt movements and try to move away. Under no circumstances should people try to feed, touch, pet, ride, or chase wild dolphins.

GUIDELINES

The following guidelines for watching and interacting with dolphins have been assembled from others used around the world.[5] They are more conservative than some, more lenient than others, and should not be followed instead of local laws. *These guidelines are not meant to condone interactions with cetaceans in any way. The most responsible way to watch cetaceans is always from a distance, and most guidelines recommend against swimming with or interacting with cetaceans for the safety of all involved.* Nonetheless, increasing numbers of people are engaging in close and in-water interaction with cetaceans. These guidelines are only very general and need to be specifically tailored in response to local people, animals, environments, and laws. For more detailed information, the NMFS Web site on marine mammal viewing can be accessed at: http://www.nmfs.noaa.gov/prot_res/MMWatch/MMViewing.html. Also, Carole Carlson's excellent review, "Whale Watch Guidelines and Regulations Around the World," can be viewed at: http://www.iwcoffice.org/SCWEB/screp2001.htm #Whalewatching.

- Never offer fish or other food items to dolphins. Never introduce inorganic items (or rubbish of any type) to dolphins.

- When in a boat, do not approach dolphins more closely than 100 yards (more if other boats are present)—dolphins will approach you if they want to be closer.

- If dolphins are near the boat, avoid making any abrupt changes of speed or course and reduce speed gradually to a "no wake" speed. The engine may also be put in neutral but only put back into gear when dolphins are at a distance. Never move between, scatter, or separate dolphins. It is particularly important to stay clear of mothers and their calves and from all dolphins who appear to be resting, feeding, or mating.

- Never pursue or chase dolphins, whether by boat or when swimming. Also, do not approach dolphins head-on or place yourself in their path of travel. (In one study, dolphins were most likely to avoid swimmers placed in their path of travel.[6])

- Avoid areas used for resting or sheltering where dolphins may be especially vulnerable to disturbance. (For example, some dolphins use shallow areas to care for their young and to avoid predators.)

- Do not touch the dolphins. If dolphins initiate touch, people should avoid sensitive areas on dolphin body: blowhole, eyes/face, genital region, flippers, dorsal fin, flukes. Do not wrap arms or hands around, grab, or restrict dolphin movement in any way.

- Leave immediately if dolphins exhibit sexual, aggressive, or submissive behavior.

- Do not enter the water abruptly if dolphins are nearby, and avoid loud or sudden noises or movements.

- Overall, be precautionary and always err on the side of caution around dolphins and all wild animals, allowing them as much space as possible.

NOTES

1. T. G. Frohoff, "Behavioral Indicators of Stress in Odontocetes During Interactions with Humans: A Preliminary Review and Discussion" (International Whaling Commission Scientific Committee, SC/52/WW2, 2000).

2. Associated Press, Aug. 29, 2002.

3. D. Seideman, "Swimming with Trouble," *Audobon* 99 (1997): 76–82; T. R. Spradlin, A. D. Terbush, and W. S. Smullen, "NMFS Update on Human/Dolphin Interactions in the Wild," *Soundings, Magazine of the International Marine Animal Trainers Association* 23, no. 1 (1998): 25–27; A. Samuels and L. Bejder, "Habitual Interactions Between Humans and Wild Bottlenose Dolphins *(Tursiops truncatus)* Near Panama City Beach, Florida" (Report to the Marine Mammal Commission, Silver Spring, Md., 1998); K. Colborn, "Interactions Between Humans and Bottlenose Dolphins, *Tursiops truncatus*, Near Panama City, Florida" (Master's thesis, Duke University, Durham, N.C., 1999).

4. L. Bryant, "Report to Congress on Results of Feeding Wild Dolphins: 1989–1994" (NOAA/National Marine Fisheries Service, Office of Protected Resources, Silver Spring, Md., 1994).

5. M. Carwardine and E. Hoyt, eds., *The Nature Company Guide to Whales, Dolphins & Porpoises* (New York: Discovery Channel and Time-Life Books, 1998); IFAW, Thethys Research Institute and Europe Conservation, "Report of the Workshop on the Scientific Aspects of Managing Whale Watching" (Montecastello di Vibio, Italy, 1995); T. G. Frohoff and K. M. Dudzinski, "Odontocete-Human Interactions: Practical Management," in *Viewing Marine Mammals in the Wild: A Workshop to Discuss Responsible Guidelines and Regulations for Minimizing Disturbance* (Vancouver, B.C., November 28, 2001); T. G. Frohoff, D. L. Herzing, and M. Santos, report from "Human Interactions with Small Cetaceans: Research and Management" (workshop convened at the 11th Biennial Conference on the Biology of Marine Mammals, Orlando, Fla., December 1995); T. R. Spradlin et al., report from "Viewing Marine Mammals in the Wild: A Workshop to Discuss Responsible Guidelines and Regulations for Minimizing Disturbance" (convened at the 14th Biennial Conference on the Biology of Marine Mammals, Vancouver, B.C., November 28, 2001).

6. R. Constantine, "Increased Avoidance of Swimmers by Wild Bottlenose Dolphins *(Tursiops truncatus)* Due to Long-Term Exposure to Swim-with-Dolphin Tourism," *Marine Mammal Science* 17, no. 4 (2001): 689–702.

Wild Dolphins,
People, and Food

Mark Orams

To share food is one of the most common ways that people show kindness, hospitality, and generosity. However, when people give food to dolphins or any wild animals, they put the animals at great risk. Dolphins and other wildlife can become dependent on "hand-outs" and may lose their ability to forage for themselves; they may become poisoned or ill from spoiled or inappropriate food; and sometimes they can even become aggressive toward each other and with people.[1]

Dolphins are ingenious and adaptable—especially when it comes to their relationships with humans. They are naturally curious and are very good at figuring out how to turn a new situation to their advantage, including obtaining food for themselves. There are many stories that illustrate this, including several from Moreton Bay, Queensland, Australia, where researchers have been studying dolphins for many years. One such story dates back to the nineteenth century.[2] Early European explorers to the area watched Aboriginal men wade into the water signing and chanting, slapping their spears on the surface. This rhythmic slapping appeared to be a signal that dolphins in the bay recognized, and soon dolphins would appear, herding large schools of small fish (probably mullet) into the shallows next to the Aborigines. Swimming in ever tightening circles, the dolphins bunched the fish schools so that the Aborigines could catch them easily with their spears and small hand nets. In return for their assistance, the Aborigines would hand-feed some of the catch to the dolphins. This cooperative fishing arrangement was said to have existed for many human and dolphin generations—learned and passed on over hundreds of years. This kind of "food sharing" is not unique to Australian history: similar dolphin-assisted fishing (or perhaps, from the dolphins' perspective, human-assisted fishing) has been reported in South America, Africa, and the Mediterranean.[3]

Another account, from more recent times, illustrates how dolphins continue to find ways of obtaining food by taking advantage of our activities. Prawn or shrimp trawling is common in Moreton Bay. Fishermen use a net that they drag across the bottom to catch these crustaceans; they also catch many bottom-dwelling fish. This net has an opening at the bottom end that is tied while trawling; after completing the trawl, when the net has been hauled up onto the boat, it is untied, which dumps the catch onto the deck of the boat for sorting. Fishermen have told me about how they began to have problems, finding that when they hauled the nets up they were already untied. Naturally, all the catch had been lost, and, apparently, a number of frustrated skippers (after several hours of trawling) accused deckhands of poor knot-tying skills. Of course, what they eventually realized was that the dolphins had learned to untie the knots! In doing so, the dolphins ended up with a "free feed." You can imagine the contest that then resulted as fishermen tried to come up with new ways of tying their nets and the dolphins continued to try and figure out how to untie them. Unfortunately, this story also illustrates the dangers for dolphins. Many dolphins are caught in trawling operations and are eventually drowned or crushed. Similarly, in a number of places, dolphins have learned that our recreational fishing from boats, docks, and piers can also provide a source of food. Many dolphins have been seen with hooks, lines, and other fishing gear stuck in their mouths and around their bodies as a result of these interactions.

Thus dolphins using human activities as a source of food are at significant risk of harm. This is also true where dolphins that are deliberately "provisioned" with food have become a tourism attraction. Monkey Mia, in Western Australia, is probably the most famous location where this occurs; people have been feeding dolphins there since the 1960s.[4] Monkey Mia is a very small town, ten hours' drive north of the closest city (Perth), and yet hundreds of thousands of people make the trip there each year just for the chance to hand-feed fish to bottlenose dolphins. Tourists are carefully managed at Monkey Mia: people are only allowed to wade into the water to their knees, and only a few people are selected by the rangers to actually feed the dolphins. Unfortunately, despite these strict controls, the feeding activities still place these dolphins at risk. Researchers have shown that dolphins have become dependent on the human-provided fish and that young dolphins have not learned to forage for themselves.[5] In one dramatic incident, a young calf was attacked and killed by a shark while his mother was distracted—begging for food close to the beach.[6]

At other places where humans feed wild dolphins, the risks are even greater. These situations are uncontrolled, usually illegal, and cause significant problems for the dolphins and the people involved. In some areas in the eastern United States such as Florida, it has become quite common for people to (illegally) feed dolphins from their boats and even from fishing piers.[7] Dolphin-feeding cruises (which are also against the law) have even been promoted by a number of boat charter operators. There have been numerous reports of people being rammed and bitten and of dolphins being injured as a result of collisions with boats and propellers.[8] Some people have even deliberately hooked dolphins and tried to "land them," resulting in painful injuries to the dolphin.

Dolphins, like all species of wildlife, are susceptible to human disturbance. This is particularly so where food is involved. The provision of food to wildlife immediately alters the natural foraging patterns of those animals. This is especially true with regard to dolphins. A great deal of a dolphin's time is spent looking for, catching, and consuming food. As a result, many dolphin species have developed cooperative hunting techniques. Their interaction with one another and with other marine creatures when fishing is fascinating, and it plays a major role in their social lives. Thus, when an artificial food source is introduced—such as when people are deliberately feeding wild dolphins—then dolphins spend much less time doing these things. Typically they travel less and interact with other dolphins and other marine animals less. Of greater concern is that, over time, they tend to lose their hunting skills—they become dependent and lose their "wildness"—and isn't that what helps to make them so special? It is clear that feeding dolphins in the wild is an activity with many risks. The only time that feeding dolphins in the wild can be justified is when the health and long-term survival of specific individuals could be compromised and when carefully controlled supplemental food may assist in their recovery.

Dolphins are supremely adaptable animals. Part of this ability to adapt has seen many dolphins learn to live in association with people. This has resulted both in wonderful experiences and, unfortunately, in tragedies. In many cases, dolphins have learned to use the activities of humans as an aid to their lives. Dolphins have utilized us (as we have them) as sources of entertainment, as objects of curiosity, as creatures to socialize with, and, perhaps, even as "friends." As sophisticated and often cooperative hunters, it is not surprising that they have also learned to utilize us, and our activities, as a source of food. Several years ago I watched, fascinated, as a young dolphin caught a live mullet, brought it into shore, and placed it gently at a child's feet! What does this mean? Perhaps only the dolphins really know. However, for dolphins to be protected from harm, our interactions with them must always be conducted with respect, with care, and with a knowledge and appreciation for the implications of our actions.

NOTES

1. M. B. Orams, "Feeding Wildlife as a Tourism Attraction: Issues and Impacts," *Tourism Management* 23 no. 3 (2002): 281–93.

2. J. K. E. Fairholme, "The Blacks of Moreton Bay and the Porpoises," *Proceedings of the Zoological Society of London* 24 (1856): 353–54.

3. C. Lockyer, "Review of Incidents Involving Wild, Sociable Dolphins, Worldwide," in *The Bottlenose Dolphin*, ed. S. Leatherwood and R. R. Reeves (San Diego, Calif.: Academic Press, 1990).

4. M. B. Orams, "Historical Accounts of Human-Dolphin Interaction and Recent Developments in Wild Dolphin–Based Tourism in Australasia," *Tourism Management* 18, no. 5 (1997): 317–26.

5. J. Mann, R. C. Connor, L. M. Barre, and M. R. Heithaus "Female Reproductive Success in Bottlenose Dolphins (*Tursiops* sp.): Life History, Habitat, Provisioning, and Group Size Effects," *Behavioural Ecology* 11 (2000): 210–19; B. Wilson, "Review of Dolphin Management at Monkey Mia" (unpublished report to the Department of Conservation and Land Management, Perth, Western Australia, 1994).

6. J. Mann, and H. Barnett, "Lethal Tiger Shark (*Galeocerdo cuvier)* Attack on Bottlenose Dolphin (*Tursiops* sp.) Calf: Defense and Reactions by the Mother," *Marine Mammal Science* 15 (1999): 568–75.

7. A. Samuels, L. Bejder, and S. Heinrich, "A Review of the Literature Pertaining to Swimming with Wild Dolphins" (unpublished report to the Marine Mammal Commission [Contract No. T 74463123], Bethesda, Md., 2000); Anonymous, "Don't Feed the Dolphins—and That's Official," *New Scientist* 140 (1993): 8.

8. L. Bryant, "Report to Congress on Results of Feeding Wild Dolphins: 1989–1994" (National Marine Fisheries Service, Office of Protected Resources, Silver Spring, Md., 1994); D. Seideman "Swimming with Trouble," *Audubon* 99 (1997): 76–82.

Conservation Issues
Facing Small Cetaceans

Sharon Young

Many people are aware that the populations of most species of large whales were dramatically reduced by commercial whaling, but few are aware that some of the greatest conservation concerns actually face cetaceans that are much smaller in size. Small cetaceans such as dolphins and porpoises are often equally endangered but receive little attention. Their plight points to many concerns facing the oceans today.

INTENTIONAL KILLING

In earlier centuries, it was common for native peoples to view dolphins as fish, and they deliberately hunted them to sustain their villages. Although many people now find this abhorrent, there are places in the world where this practice continues to occur, now for commercial gain.

In countries such as Peru, Sri Lanka, and the Philippines, dolphins that used to be accidentally caught in nets are now deliberately targeted and sold as meat in markets or used for bait in crab fisheries. In the Faeroe Islands in the North Atlantic, people who are affluent enough to drive Volvos also deliberately drive more than a thousand pilot whales to their deaths each year in shallow coastal waters. This, and other examples of humans deliberately driving small cetaceans into shallow waters in order to slaughter them, result in the deaths of countless animals each year. In Japan, these "drive fisheries" for various dolphin species have resulted in hundreds of thousands being killed between the 1940s and the 1990s. Today, the populations are so reduced that fishermen cannot find enough dolphins to fill quotas set by their government, and the International Whaling Commission has recommended that this fishery be halted. The practice still continues, however, with dolphins being slaughtered or sold to aquariums and zoos for exhibition. In fact, eight Japanese states allow dolphin hunts with annual quotas of 20,000 dolphins.

FISHERY BYCATCH OF DOLPHINS

Species that are unintentionally caught by fisheries are referred to as bycatch. The best-known example of marine mammals as bycatch is that of dolphins in the Pacific; they were caught by tuna boats whose operators deliberately set their nets around dolphins in order to catch the tuna that often swam beneath them. As recently as 1991, it was estimated that 11,000 northern right whale dolphins died each year in this manner. Populations of eastern spinner dolphins were reduced to less than half of their original numbers. Changes in U.S. law that prevented fishermen from importing tuna caught in this manner resulted in dramatically reduced mortality of dolphins from a variety of species, although it is estimated that as many as 5,000 per year are still being killed. As of this writing, courts have become involved in determining whether or not it will be legal to sell tuna in the United States if it is caught in a manner that deliberately targets dolphins.

Unintentional entanglement is a serious conservation concern for a variety of species. There are only approximately 300 North Atlantic right whales remaining. Many of them are entangled each year in lines used to set lobster traps, and they are dying at a rate that jeopardizes their species' existence. Dolphins and porpoises also face this crisis. Fishermen along the U.S. East Coast have been forced to modify fishing practices to reduce unsustainable levels of mortality in species such as bottlenose dolphins in the mid-Atlantic region and harbor porpoise in the Northeast. Perhaps the most extreme problem is faced by the vaquita, a small porpoise that lives only in the northern Gulf of California in Mexico. This secretive species is difficult to study, and its population is estimated at only around 500 animals. They are caught in gillnets at a rate that is considered unsustainable, and scientists fear for the continued survival of this species.

THREATS TO HABITAT

In many ways, the oceans are our last frontier on the planet. They are vast areas about which we know little. Much as terrestrial frontiers were altered and species endangered by burgeoning human populations, many ocean dwellers are beginning to face similar sorts of habitat loss and degradation that threaten their populations.

Climatic change occurs with or without humans, but our production of greenhouse gases may hasten the warming of the planet. Scientists have been able to measure shrinking icecaps at the poles and changes in the extent of pack ice in cold polar waters. These changes cause a shift in ocean temperature and in the currents that flow through the oceans. The result is a disruption in the food chain that can affect marine mammals. For example, climate-induced changes in the movement of currents off the U.S. West Coast and the Barents Sea in the Russian arctic have resulted in the populations of some fish species increasing while other species decline. This may favor marine animals that eat the increasing fish species but harm marine animals that rely on the declining fish species. In addition to broad-scale changes in ocean habitats, other, smaller-scale changes may affect small whales.

Chemical pollutants, which wash down from inland rivers or leach from coastal septic systems, degrade the coastal habitats of dolphins and porpoises. In 1990, large numbers of dead striped dolphins were found along the Mediterranean coastlines of Spain, France, and northern Italy. There was a similar die-off of bottlenose dolphins on the U.S. mid-Atlantic coast in the late 1980s. In both cases, the dolphins had high levels of contaminants in their tissues. There has been speculation that pollutants may have played a prominent role in compromising their immune systems and making them more vulnerable to diseases or environmental stressors that might otherwise not have harmed them.

Acoustic pollution is also a concern. The use of loud sounds such as that discussed elsewhere in this book may injure or displace marine mammals. Sounds from oil exploration, underground blasting, and naval ordnance exercises also may harm animals. Adverse effects may occur without our being aware of them, because many species such as beaked whales and oceanic dolphins live too far from shore for us to study them with any ease. But we do know that loud sounds can have a significant effect. As coastal salmon aquaculture farmers try to drive away seals that seek to eat the penned fish, they sometimes use powerful acoustic devices that give off loud pulses. These acoustic harassment devices have been found in some experiments to drive porpoises as far as three kilometers away from the sites. Displacing animals into less optimal habitats may jeopardize their nutritional health and reproductive success.

Some dolphins live in river systems and may be vulnerable to all of the above problems and a host of others. This is the case for the Baiji dolphins in the Yangtze River. Dams along the river have significantly reduced their range, crowding them into smaller and less desirable areas where they are also vulnerable to being caught in fishing gear. With only a couple of hundred animals remaining, there is grave concern over the species' survival.

LESSONS LEARNED

Southern resident orcas that live along the coast of Washington State in the United States and British Columbia, Canada, illustrate many of these problems. Populations were originally reduced when they were captured by oceanariums that wished to display them to the public. After this hunting ceased, their numbers increased, but today they are declining once more, with fewer than 100 remaining in the population. They have been shot by fishermen, fearing competition. They are at risk from oil spills in the shipping lanes. The chemical toxin PCB has been found in their bodies at levels that are 500 times the levels in the average human. These orcas rely heavily on eating Chinook salmon, which are themselves endangered in many areas as a result of the loss of riverine habitats to development and dams. Orcas are targeted by the whale-watch industry, whose many boats may surround or pursue them for long periods of time (200 boats surrounding five or six whales have been observed at a time). Scientists fear that this may stress the animals and disrupt normal feeding and breeding behavior. All of these factors may contribute to their low reproductive rates and the ongoing disappearance of members of the southern resident community.

What we can learn from the plight of smaller cetaceans is that our activities may compound natural ecosystem fluctuations, putting populations at risk of extinction. There are a number of ways in which each of us can help improve the prospects for dolphins and whales. For example, join or volunteer for nonprofit groups who support your desire to protect cetaceans. When you hear about proposals that would weaken laws or regulations that protect dolphins and whales, take a moment and write a letter to your elected representatives. A few well-chosen words can make a big difference. If you own a boat, operate it carefully to avoid harassing animals. Recognizing the vulnerability of ocean animals to acoustic and chemical pollutants is vital. Reducing sources of coastal pollution and regulating fisheries that may intentionally or incidentally kill them can significantly reduce risk to these animals. In the end, their ocean world is vital to us as well. A healthy ocean for marine mammals means a healthier planet for all of us.

HELPFUL REFERENCES

Evans, G. H., and J. A. Raga. 2001. *Marine Mammals Biology and Conservation*. Hingham, Mass.: Kluwer Academic/Plenum Publishers.

International Whaling Commission Web site: http://www.iwcoffice.org/iwc.htm.

Marine Mammal Commission. 2001. *Annual Report to Congress*. Bethesda, Md.

Perrin, W. F., B. Würsig, and H. G. M. Thewissen. 2002. *Encyclopedia of Marine Mammals*. San Diego, Calif.: Academic Press.

Twiss, J. R., and R. R. Reeves. 1999. *Conservation and Management of Marine Mammals*. Melbourne, Australia: Melbourne University Press.

Organizations

American Cetacean Society
P.O. Box 1391
San Pedro, CA 90733
www.acsonline.org

Blue Voice.org
Contact: Hardy Jones
1252 B Street
Petaluma, CA 94952
www.bluevoice.org

Centre d'Etudes Hydrobiologiques
Contact: Monica Müller, Ph.D.
108 Avenue du Puig del Mas
66650 Banyuls-sur-Mer, France

Cetacean Society International
Contact: William W. Rossiter,
President
P.O. Box 953
Georgetown, CT 06829
www.csiwhalesalive.org

Coastal Marine Research Group
Contact: Mark Orams, Ph.D., Director
Massey University at Albany
Private Bag 102-904
North Shore MSC, Auckland,
New Zealand
http://cmrg.massey.ac.nz/

Conservación de Mamíferos Marinos
de México, A.C.
(COMARINO)
Xochicalco 149-A
Col Narvarte
Mexico, D.F. C.P. 03020

Dalhousie University
Department of Biology
Contact: Lindy Weilgart, Ph.D.,
Research Associate
Halifax, Nova Scotia B3H 4J2, Canada
http://is.dal.ca/~whitelab

Dolphin Communication Project
Contact: Kathleen M. Dudzinski,
Ph.D., Director
3600 South Harbor Blvd., #429
Oxnard, CA 93035
www.dolphincommunicationproject.org

Dusky Dolphin Project & Hawai'i
Marine Mammal Consortium
Contact: Suzanne Yin, Researcher
1483 Sutter Street #1107
San Francisco, CA 94109
www.hmmc.org

Earth Island Institute, International
Marine Mammal Project
300 Broadway #28
San Francisco, CA 94133
www.EarthIsland.org/immp

Ethologists for the Ethical Treatment
of Animals and Citizens for Responsible Animal Behavior Studies
Marc Bekoff, Ph.D., and Jane Goodall,
Ph.D., Co-founders
www.EthologicalEthics.org

Fundacion Dominicana de Estudios
Marinos, Inc.
Contact: Idelisa Bonnelly de Calventi,
Presidenta
Calle Sócrates Nolasco No. 6,
Apto. 401
Residencial Carla Pamela,
Ensanche Naco
Santo Domingo,
República Dominicana
Email: fundemar@codetel.net.do

Humane Society of the United States
Contact: Naomi A. Rose, Ph.D.,
Marine Mammal Scientist
Wildlife and Habitat Protection
2100 L Street, NW
Washington, D.C. 20037
www.hsus.org

Institute of Marine Life Sciences
Contact: Bernd Würsig, Ph.D.
Texas A&M University
4700 Avenue U, Bldg. 303
Galveston, TX 77551

International Dolphin Watch
10 Melton Road
North Ferriby
East Yorkshire HU14 3ET
England
www.idw.org

Interspecies Inc.
Contact: Jim Nollman, Founder
www.Interspecies.com

John C. Lilly Research Institute, Inc.
P.O. Box 791695
Paia, HI 96779-1695
www.johnclilly.com

National Marine Fisheries Service
Office of Protected Resources
Provides program oversight, national
policy direction, and guidance on the
conservation of those marine mammals
and endangered species, and their habitats, under the jurisdiction of the Sec-
retary of Commerce in the United
States.
www.nmfs.noaa.gov/prot_res
/overview/mm.html

MARMAM
An edited e-mail discussion list that
focuses on marine mammal research
and conservation, run through the
University of Victoria. The list was
established in August 1993 specifically
for marine mammal researchers and
graduate students as well as those
actively involved in marine mammal
conservation and management.
http://is.dal.ca/~whitelab/marmam.htm

Ocean Futures Society
325 Chapala Street
Santa Barbara, CA 93101
www.oceanfutures.org

Ocean Mammal Institute
P.O. Box 14422
Reading, PA 19612
www.oceanmammalinst.org

Online Zoologists
www.onlinezoologists.com/topics/
marspec/

Orca Conservancy
71 Little Creek Road
Friday Harbor, WA 98250
www.orcaconservancy.org

OrcaLab/Pacific Orca Society
Contacts: Dr. Paul Spong and
Helena Symonds
P.O. Box 510
Alert Bay, British Columbia
V0N 1A0, Canada
www.orcalab.org

Orca Network
2403 S. North Bluff Road
Greenbank, WA 98253
www.orcanetwork.org

Project Interlock
Contact: Wade Doak, Founder
Box 20
Whangarei, New Zealand
www.wadedoak.com

Society for Marine Mammalogy
The objectives of the Society are to: (1)
evaluate and promote the educational,
scientific, and managerial advancement
of marine mammal science; (2) gather
and disseminate to members of the
Society, to the public, and to public
and private institutions scientific,
technical, and management information through publications and meetings; and (3) provide scientific information, as required, on matters related to
the conservation and management of
marine mammal resources.
www.marinemammalogy.org

Swiss Working Group for the Protection of Marine Mammals (ASMS)
P.O. Box 30
CH-8820 Wädenswill, Switzerland
www.asms-swiss.ch

TerraMar Research
Contact: Toni G. Frohoff, Ph.D.,
Director
321 High School Road NE,
PMB 374
Bainbridge Island, WA 98110
www.TerraMarResearch.org

Tethys Research Institute
Contact: Giovanni Bearzi,
Contract Professor of Cetacean
Conservation
University of Venice
c/o Venice Natural History Museum
Santa Croce 1730
30135 Venezia, Italy
www.tethys.org

Whale and Dolphin Conservation
Society
Brookfield House
38 St. Paul Street
Chippenham, Wiltshire SN15 1LY,
England
www.wdcs.org

The Whale Museum
62 First Street North
P.O. Box 945
Friday Harbor, WA 98250
www.whale-museum.org

Whale Stewardship Project
Contact: Cathy Kinsman, Executive
Director
P.O. Box 36101
Halifax, Nova Scotia B3J 3S9, Canada
www.whalestewardship.org

Wild Dolphin Project
P.O. Box 8436
Jupiter, FL 33468
www.wilddolphinproject.com

Recommended Reading

Bekoff, Marc, ed. 2000. *The Smile of a Dolphin: Remarkable Accounts of Animal Emotions.* New York: Discovery Books.

Carwadine, Mark. 1995. *Whales, Dolphins, and Porpoises: The Visual Guide to All the World's Cetaceans.* Eyewitness Handbooks. New York: Dorling Kindersley.

Carwadine, M., and Erich Hoyt, eds. 1998. *The Nature Company Guide to Whales, Dolphins & Porpoises.* New York: Discovery Channel and Time-Life Books.

Collet, Anne. 2000. *Swimming with Giants: My Encounters with Whales, Dolphins, and Seals.* Minneapolis, Minn.: Milkweed Editions.

Dierauf, Leslie A., and Frances M. D. Gulland, eds. 2001. *The CRC Handbook of Marine Mammal Medicine,* 2d edition. London: CRC Press.

Doak, Wade. 1988. *Encounters with Whales and Dolphins.* Dobbs Ferry, N.Y.: Sheridan House.

Dudzinski, Kathleen. 2000. *Meeting Dolphins: My Adventures in the Sea.* Washington, D.C.: National Geographic Society. (Ages 9–12)

Hoelzel, A. Rus, ed. 2002. *Marine Mammal Biology: An Evolutionary Approach.* Oxford: Blackwell Science.

Hoyt, Erich. 2001. *Whale Watching 2001: Worldwide Tourism Numbers, Expenditures, and Expanding Socioeconomic Benefits.* Yarmouth Port, Mass.: International Fund for Animal Welfare.

Leatherwood, Stephen, and Randall R. Reeves; photographs by Larry Foster. 1992. *The Sierra Club Handbook of Whales and Dolphins.* San Francisco: Sierra Club Books.

Lilly, John C. 1975. *Lilly on Dolphins: Humans of the Sea.* Garden City, N.Y.: Anchor Books.

Mann, Janet, Richard C. Connor, Peter L. Tyack, and Hal Whitehead, eds. 2000. *Cetacean Societies: Field Studies of Dolphins and Whales.* Chicago: University of Chicago Press.

McIntyre, Joan. 1974. *Mind in the Waters: A Book to Celebrate the Consciousness of Whales and Dolphins.* New York and San Francisco: Charles Scribner's Sons and Sierra Club Books.

Nollman, Jim. 1999. *The Charged Border: Where Whales and Humans Meet.* New York: Henry Holt.

Perrin, William F., Bernd Würsig, and J. G. M. Thewissen, eds. 2001. *Encyclopedia of Marine Mammals.* San Diego, Calif.: Academic Press.

Pryor, Karen, and Kenneth S. Norris, eds. 1991. *Dolphin Societies: Discoveries and Puzzles.* Berkeley: University of California Press.

Reeves, Randall R., Brent S. Stewart, Phillip J. Clapham, and James A. Powell; illustrated by Pieter A. Folkens. 2002. *Marine Mammals of the World.* New York: Alfred A. Knopf.

Reynolds III, John E., and Sentiel A. Rommel. 1999. *Biology of Marine Mammals.* Washington, D.C.: Smithsonian Institution Press.

Simmonds, Mark P., and Judith D. Hutchinson, eds. 1991. *The Conservation of Whales and Dolphins: Science and Practice.* New York: John Wiley & Sons.

Smolker, Rachel. 2001. *To Touch a Wild Dolphin.* New York: Nan A. Talese/ Doubleday.

Twiss Jr., John R., and Randall R. Reeves. 1999. *Conservation and Management of Marine Mammals.* Washington, D.C.: Smithsonian Institution Press.

The Whale Watching Web. An online cetacean bibliography, audiography, and videography compiled by Trisha Lamb Feuerstein: www.physics.helsinki.fi/whale/literature/fic_main.html.

ABOUT THE CONTRIBUTORS

DIANE ACKERMAN, PH.D., is a poet, essayist, and naturalist. She is the author of numerous books, including *A Natural History of the Senses; The Moon by Whale Light and Other Adventures Among Bats, Penguins, Crocodilians, and Whales; Deep Play;* and *Cultivating Delight: A Natural History of My Garden.* Ackerman has received many awards, including the John Burroughs Nature Award and the Academy of American Poets' Lavan Poetry Prize.

GIOVANNI BEARZI, Pew Marine Conservation Fellow, has carried out cetacean surveys in the Mediterranean Sea since he could enact free will, focusing on the behavioral ecology of coastal dolphins. After having founded and directed for a decade a dolphin research and conservation program in Croatia—which was awarded the Henry Ford European Conservation Award as best European project overall—Giovanni moved to Venice, Italy, where he created a dolphin research and information center. He is President of the Tethys Research Institute, member of the IUCN Cetacean Specialist Group, and contract professor of cetacean conservation at the Faculty of Environmental Sciences, University of Venice, Italy.

MARC BEKOFF, PH.D., is a Fellow of the Animal Behavior Society and a former Guggenheim Fellow. He is author or editor of many books, including *Strolling with Our Kin; The Smile of a Dolphin: Remarkable Accounts of Animal Emotions; Minding Animals: Awareness, Emotions, and Heart;* and (with Jane Goodall) *The Ten Trusts: What We Must Do to Care for the Animals We Love.* He and Jane co-founded Ethologists for the Ethical Treatment of Animals: Citizens for Responsible Animal Behavior Studies (www.ethologicalethics.org).

LEIGH CALVEZ has worked as both a naturalist and a scientist studying whales and dolphins in Massachusetts, Hawaii, and Washington State over the last ten years. As a nature writer, her interests have taken her to Canada to watch polar bears, to the Great Bear Rainforest in British Columbia to search for spirit bears, and to India to find tigers. Her work has been published in *Ocean Realm, The Ecologist, Christian Science Monitor, Seattle Post-Intelligencer,* and *Seattle Times.*

ROCHELLE CONSTANTINE, PH.D., has been studying whales and dolphins for eleven years with a particular interest in the effects of tourism. With careful management, good science, and teaching people respect for these interesting mammals, she hopes that we will minimize our impact upon the dolphins as they go about their daily lives.

355

HORACE DOBBS, PH.D., a "Renaissance Man," is a multi-discipline scientist who has published papers on atomic physics, chemistry, pharmacology, and medicine. He was elected a Fellow of the Royal Society of Medicine and has won numerous international awards for his visionary books and television films. He speaks at hospitals, schools, colleges, and universities around the world.

KATHLEEN DUDZINSKI, PH.D., has been studying dolphin behavior and communication primarily in the Bahamas and Japan but also in Belize, the Gulf of Mexico, Patagonia, and Europe since 1990. She founded and directs the Dolphin Communication Project, which focuses on dolphin research, conservation, and education programs. Kathleen also advises students at University of Southern Mississippi as an adjunct faculty member in Psychology.

CATHY ENGLEHART, DC, is a chiropractor in Seattle, Washington. She has studied with the Upledger Institute and currently is pursuing studies in Visionary Craniosacral work with the Milne Institute. Her work involves perception of body structures not simply as muscles and bones but as aspects of consciousness.

TONI FROHOFF, PH.D., has been studying marine mammal behavior and the effects of human activities (such as swim-with-the-dolphin programs) on dolphins in captivity and in the wild for almost twenty years. Her research for government and nongovernment agencies (including the Humane Society of the United States and Earth Island Institute) has contributed to the revision and implementation of legislation protecting captive and free-ranging dolphins in several countries. As Research Director for TerraMar Research, she is currently studying stress in dolphins.

HOWARD GARRETT served as editor of *Cetus*, the journal of the Whale Museum, and is the author of *New England Whales* and *Orcas in Our Midst*, a booklet depicting orcas and their dependence on salmon and healthy watersheds. A frequent speaker and writer about orcas and environmental issues, he is co-founder and president of Orca Network. He is happiest describing the astounding capabilities and social lives of orcas and the deepening bond between whales and people.

OZ GOFFMAN received his master's degree from the University of Haifa in 1997, where he studied the behavior of a solitary, sociable dolphin. In 1994, Goffman and others formed the Israel Marine Mammal Research and Assistance Center, the first organization in the Middle East concerned with educating the public about the marine mammal populations off the coast of Israel and its Arab neighbor states and dedicated to aiding stranded cetaceans. In 1999, Goffman was accepted as a Ph.D. student at the University of Haifa to continue his studies on the sociable, solitary dolphin.

LINDA HOGAN, a Chickasaw poet and essayist, is a recipient of a National Endowment for the Arts grant in fiction, a Guggenheim Fellowship, a Lannan Fellowship, and the Five Civilized Tribes Museum Playwriting Award. She has been shortlisted for the Pulitzer Prize and the National Book Critics Circle Award, and she won the American Book Award for *Seeing Through the Sun*. Her numerous books include *Calling Myself Home*, *Solar Storms*, *Power*, and, most recently, *Sightings: The Gray Whales' Mysterious Journey* (with Brenda Peterson).

ERICH HOYT is a writer, naturalist, and researcher who has gone whale and dolphin watching in some thirty countries. His more than 350 publications include fourteen books, such as *Orca: The Whale Called Killer*, *The Earth Dwellers*, *Insect Lives*, and *Creatures of the Deep*. An American-Canadian dual citizen now based in Scotland, Erich is Senior Research Associate with the Whale and Dolphin Conservation Society and Co-director of the Far East Russia Orca Project.

CATHY KINSMAN has dedicated much of her time to the study and protection of whales and dolphins in Canada and around the globe since 1990. Her love of cetaceans and the natural world inspires her to communicate an ethic of compassion and respectful coexistence through writing, filmmaking, musical composition, painting, and scientific research.

JOHN C. LILLY, M.D., an unparalleled scientific visionary and explorer, made significant contributions to psychology, brain research, computer theory, medicine, ethics, delphinology, and interspecies communication before his death in 2001. His work helped launch the global interest in dolphins and whales, providing the basis for the book and movie *Day of the Dolphin* and stimulating the enactment of the Marine Mammal Protection Act in 1972. He is largely credited as the grandfather of many modern dolphin researchers, and his work has been an inspiration to an entire generation of dolphin enthusiasts.

CHRISTINA LOCKYER, D.SC., is Senior Research Scientist at the Danish Institute for Fisheries Research, Department of Marine Ecology and Aquaculture, in Denmark. Generally, her research is related to marine mammal management issues and focuses on age determination, life history parameters, growth, feeding and reproductive energetics, and population structure.

ASHLEY MONTAGU, PH.D., was a British-American anthropologist world renowned for his popular books such as *The Nature of Human Aggression*, *Growing Young*, *Science and Creationism*, *Man's Most Dangerous Myth: The Fallacy of Race*, and *The Natural Superiority of Women*.

SY MONTGOMERY is a columnist, documentary scriptwriter, and commentator of National Public Radio's "Living on Earth." She is the author of eight books, including *Encantado: Pink Dolphin of the Amazon* (for children), her most recent book, and *Search for the Golden Moon Bear* (for adults).

MONICA MÜLLER, PH.D., is an ethologist specializing in the behavior of solitary and sociable dolphins in France, Spain, Ireland, New Zealand, and Australia, where she has been studying during a five-year doctoral research period. The focus of her studies has been to find out why wild dolphins become solitary and interact with humans and how human-dolphin interactions can be managed.

JIM NOLLMAN is the founder of Interspecies Inc. He has conducted communication research with many species of whales. His books include *The Beluga Café*, *The Charged Border, Dolphin Dreamtime, Why We Garden*, and *The Man Who Talks to Whales*.

MARK ORAMS, PH.D., is currently the director of the Coastal-Marine Research Group at Massey University at Albany in New Zealand. Much of his research focuses on the impacts of tourism on dolphin biology and behavior, but he also conducts work on the wider issues surrounding the management of human impacts on marine resources. He has authored or coauthored three books and over twenty scientific papers on the subject.

BRENDA PETERSON is a nature writer and novelist whose acclaimed nonfiction works include *Living Water; Singing to the Sound: Visions of Nature, Animals, and Spirit;* and *Build Me an Ark: A Life with Animals*. She is also a co-editor of the anthology *Intimate Nature: The Bond Between Women and Animals*. *Duck and Cover*, one of her three novels, was named a *New York Times* "Notable Book of the Year." Her latest nonfiction work is *Sightings: The Gray Whales' Mysterious Journey* (with Linda Hogan).

NAOMI A. ROSE, PH.D., has been the marine mammal biologist for the Humane Society of the United States (HSUS) since 1993. She works on marine mammal protection issues—including legislation, education, litigation, and investigation—both nationally and internationally. She is a member of the International Whaling Commission Scientific Committee and represents the HSUS on several coalitions working to protect marine mammals from the negative impacts of human activities.

BILL ROSSITER first focused his love of nature on whales and dolphins in 1973, and with his wife, Mia, by 1980 was investigating the interactive nature of cetaceans with their still-charmed Zodiac, "Morfil." Hundreds of "curious" individuals from thirteen species have enabled decades of benign and mutual exploration into the cognitive nature of whales and dolphins, and they have empowered Bill's advocacy work through Catacean Society International. His "other" career was forty years as a now-retired pilot for the U.S. Air Force and United Airlines.

MARCOS SANTOS is a biologist who received his master's degree in ecology in 1999 studying cetacean mortality. Currently, he is a doctoral student at the Universidade de São Paulo, where he is conducting a long-term study of a marine tucuxi dolphin population in southeastern Brazil. In 1996 he published a book on cetaceans for twelve- to sixteen-year-old students.

BETSY SMITH, PH.D., is a founding professor at Florida International University in Miami. Her research with neurologically impaired and autistic individuals led to her development of dolphin-assisted therapy. In recognition of this, Betsy was awarded a Best Social Inventions Award in 1990 from the Institute for Social Inventions, London, and was featured in the Rolex 1990 Awards for Enterprise. Although ethical considerations have caused her to terminate her research with dolphins, she continues investigation into the areas of domestic animal and eco-therapies. She currently teaches a graduate animal-assisted treatment course that enrolls students from around the world.

RACHEL SMOLKER, PH.D., co-founded the Monkey Mia Dolphin Research Project in 1982, which continues to produce groundbreaking insights into virtually every aspect of dolphin life. She has participated in other studies of dolphins and whales all over the world. She is currently a research associate at the University of Vermont and maintains an affiliation with the Museum of Zoology at the University of Michigan, where she completed her doctorate.

PAUL SPONG, PH.D., and HELENA SYMONDS are the directors of OrcaLab, a land-based whale research station on Hanson Island in British Columbia, Canada. Their work focuses on the life history of the northern resident community of British Columbia orcas. Helena's special interest is in orca acoustics; Paul's is in the development of technology that allows nonintrusive access to the lives of whales.

JOANA MCINTYRE VARAWA is the author of the books *Mind in the Waters*, *The Delicate Art of Whale Watching*, and *Changes in Latitude*. She founded and directed Project Jonah, the first organization to campaign internationally for a world moratorium on commercial whaling, and originated the international campaign to make the wearing of wild furs unfashionable. She lives on an island in the Pacific and is in daily contact with the astonishing and marvelous realities of nature.

LINDY WEILGART, PH.D., has been working on the acoustic communication and behavior in whales (particularly sperm whales) since 1982. Her master's and doctoral work as well as her post-doctoral fellowship were all devoted to this subject area. More recently, she has become involved in raising awareness about the dangers of undersea noise to marine mammals.

PAT WEYER, PH.D., is a Seattle multimedia artist and teacher working primarily in glass who recently completed her doctoral research on delphinology and the visual arts. With glass sculpture the artist has created a mythical watery world that reflects the actual dolphin habitats in which she has worked while serving as a volunteer artist for The Wild Dolphin Project. As a Pilchuck scholar, Pat has earned three nominations for the Corning Prize and the Alice Rooney Women in Glass Award for her glass forms.

BEN WHITE is the international coordinator for the Animal Welfare Institute. He handles all whale, dolphin, and forest issues for the AWI and is its representative to the International Whaling Commission, CITES, and IATTC and is noted for his direct action efforts.

RICHARD WILBUR has written numerous books, including *New and Collected Poems*, which won the Pulitzer Prize; *The Mind-Reader: New Poems; Advice to a Prophet and Other Poems;* and *Things of This World,* for which he received the Pulitzer Prize and the National Book Award. He served as U.S. Poet Laureate in 1987–88.

BERND WÜRSIG, PH.D., AND MELANY WÜRSIG, currently with Muritai Maui, Kaikoura, New Zealand, have been studying dolphins and whales since 1972. Bernd teaches and does research at Texas A&M University; Melany is a primary school teacher in Houston.

SUZANNE YIN has been studying the behavioral ecology of whales and dolphins for over ten years. Her work has taken her to Maine, Hawaii, Alaska, Costa Rica, New Zealand, and Japan. She received her master's degree in Wildlife and Fisheries Sciences from Texas A&M University in 1999, looking at the effects of tourism on the movement patterns and acoustic behavior of dusky dolphins.

SHARON YOUNG has worked for ten years studying whales in New England and Canada. Since 1992, she has worked for the Humane Society of the United States as a marine mammal policy analyst and advocate specializing in marine mammal interactions with commercial fisheries.

PERMISSION
ACKNOWLEDGMENTS